THE LIVING EARTH

HISTORICAL GEOLOGY

UNDERSTANDING OUR PLANET'S PAST

JON ERICKSON

FOREWORD BY PETER D. MOORE, PH.D.

☑® Facts On File, Inc.

HISTORICAL GEOLOGY
Understanding Our Planet's Past

Copyright © 2002 by Jon Erickson

Facts On File, Inc.
132 West 31st Street
New York NY 10001

Library of Congress Cataloging-in-Publication Data

Erickson, Jon, 1948–
    Historical geology : understanding our planet's past / Jon Erickson ;
foreword by Peter D. Moore.
      p. cm.—(The living earth)
    Includes bibliographical references and index.
    ISBN 0-8160-4726-X (acid-free paper)
    1. Historical geology. I. Title.

QE28.3.E73 2002
551.7—dc21   2001023055

Facts On File books are available at special discounts when purchased in bulk quantities for businesses, associations, institutions, or sales promotions. Please call our Special Sales Department in New York at 212/967-8800 or 800/322-8755.

You can find Facts On File on the World Wide Web at http://www.factsonfile.com

Text design by Cathy Rincon
Cover design by Nora Wertz
Illustrations by Jeremy Eagle and Dale Dyer © Facts On File

Printed in the United States of America

VB Hermitage 10 9 8 7 6 5 4 3 2 1

This book is printed on acid-free paper.

CONTENTS

TABLES

ACKNOWLEDGMENTS

The author thanks the Chicago Field Museum of Natural History, the National Aeronautics and Space Administration (NASA), the National Museums of Canada, the National Optical Astronomy Observatories (NOAO), the National Park Service, the U.S. Geological Survey (USGS), and the U.S. Navy for providing photographs for this book.

The author also thanks Frank K. Darmstadt, Senior Editor, and Cynthia Yazbek, Associate Editor, for their invaluable contributions to the development of this book.

FOREWORD

I f we really want to understand another person we would probably begin by examining their history, the environment in which they were raised, their experiences when they were young, any catastrophes they have had to endure. The same is true of our planet Earth. This is the only home we shall ever have and it is wise for us to understand how it functions so that we can adequately take care of it. To do this we need to know about its history.

Perhaps we have never considered just how remarkable, perhaps unique, our planet is. We exist here today because of an amazing series of geological coincidences, each of which has contributed to the life-supporting conditions we currently find around us. This book is the story of that almost incredible sequence of events that has culminated in the comforts of our Earth. We are just the right distance from our neighborhood star, the Sun, to provide the energy we need, but not in excess. The Moon is a stabilizing influence on our planet's rotation and tilt, ensuring the regularity of our seasons, our seed-time and harvest. The temperature of the Earth allows that vital material, water, to remain in a liquid state over much of the planet's surface, acting as the birth-place, the support system, and indeed the major component of life. The chemicals needed for life, such as nitrogen, phosphorus, potassium, and sulfur, were all present and available in the ancient oceans. Thus the unthinkable happened, and life emerged, soon developing the vital green pigment, chlorophyll, by means of which the energy of our Sun could be harnessed and the future of living things assured.

As you read this book you will find that the history of life and the history of the rocks are deeply and intricately connected. The rocks contain fragments and records of former living things, allowing us to reconstruct the story of their development. But those creatures have also had great impacts on sediments, the oceans, and the atmosphere of the planet. It was those first primitive photosynthetic organisms, for example, that began to pump oxygen into the early atmosphere. To them it was a waste product of energy capture, but its production so altered the history of the Earth that both the rocks and the subsequent forms of life were profoundly influenced by its presence. The development of photosynthesis can thus be regarded as the single most important step in the evolution of life. It was the turning point that was to paint the planet green. With oxygen in the atmosphere, conditions became right for massive chemical changes, such as the rusting of iron, and, more important, the scene was set for living things to invade the land.

The sheer diversity of forms that life has adopted defies the imagination. We still delight in the Earth's biodiversity, but what remains is only a tiny fraction of what lies dead and preserved beneath our feet in the rocks. The fossils of lost life-forms imply great rates of extinction, sometimes resulting from catastrophes of mind-bending magnitude, such as the asteroid collision with the Earth at the time of the demise of the dinosaurs. The recurrent spread of ice sheets at times in the Earth's history have also left their mark on the rocks and on current patterns of life, as have the lumbering movements of whole continents, tearing apart, drifting, and colliding over the crust of the planet.

In the final scene of the story set out in this book there arrives a new player—our own species. No other biological event since the development of photosynthesis has had such a major impact on the planet. The current rates of extinction may prove as disastrous for the Earth as the arrival of another asteroid. We can only hope that the appreciation of our biological and geological heritage, as set out here, will help to make the reader aware of the dangers and thus reduce the effects of our impact on this planet. Perhaps the last chapter of this book will not prove to be the final chapter after all.

—Peter D. Moore, Ph.D.

INTRODUCTION

The most fascinating field of geology is the study of our planet's past. The history of Earth is written in its rocks, and the history of life is told by its fossils. Earth history is divided into units of geologic time according to the type and abundance of fossils in the strata. The fossil record provides valuable insights into the evolution of Earth. Knowledge of the origination and extinction of species throughout geologic time is also necessary for understanding the evolution of life.

When fossils are arranged chronologically, they vary in a systematic way according to their positions in the geologic column. This observation led to one of the most important and basic principles of historical geology, whereby periods of geologic time could be identified by their distinctive fossil content. Geologists were thus able to recognize geologic time periods based on groups of organisms that were especially abundant and characteristic during a particular interval. The occurrence of certain organisms defined each period. The succession of species was the same on every continent and was never out of order. These principles became the basis for establishing the geologic time scale and ushered in the birth of modern geology.

The text chronicles the formation of Earth and its evolving life-forms, beginning with the earliest history of the planet during the Archean eon. It follows the evolution of the more complex life-forms of the Proterozoic eon. It then examines the invertebrate life-forms of the lower Paleozoic era, the early vertebrate life of the Ordovician period, and the plant life of the Silurian period. Next, it focuses on life in the sea and the first vertebrates to come onto the land during the Devonian period. It covers the evolution of the amphibians in the great coal swamps of the Carboniferous period and the evolution of the reptiles during the Permian period followed by the greatest mass extinction in Earth history.

The text continues with the evolution of the dinosaurs during the Triassic period. It explores the creatures that took flight along with the largest animals ever to roam Earth during the Jurassic period. It follows the life-forms and landforms of the Cretaceous period and the extinction of the dinosaurs. It then discusses the evolution of the mammals during the Tertiary period. Finally, it examines life during the ice ages and the present interglacial during the Quaternary period.

Students of geology and earth science will find this a valuable reference book to further their studies. Readers will enjoy this clear and easily readable text that is well illustrated with compelling photographs, wonderful drawings, and helpful tables. A comprehensive glossary is provided to define difficult terms, and a bibliography lists references for further reading. Science enthusiasts will particularly enjoy this fascinating subject and gain a better understanding of how the living Earth has evolved down through the ages.

1

PLANET EARTH
ORIGIN OF LAND AND LIFE

This chapter chronicles the formation of Earth and the creation of life. Earth was born during a time of unusual stellar formation. Only a small percentage of the stars in the Milky Way galaxy are single, medium-sized stars such as the Sun, and fewer yet have a system of orbiting planets. Therefore, the Sun with its nine planets and their satellites is considered a rarity among stars.

Earth is unique among the planets, with its roving continents and evolving life-forms. It is the only planet with a water ocean and an oxygen atmosphere. It also has a rather large moon compared with its mother planet, a pairing that still defies explanation. The atmosphere and ocean evolved during a tumultuous period of crustal formation, volcanic outgassing, and cometary degassing. Numerous giant meteorites slammed into Earth, adding unique ingredients to the boiling cauldron. Raging storms brought deluge after deluge and unimaginable electrical displays. Out of this chaos came life.

THE SOLAR SYSTEM

Far back in the very obscured past, some 12 billion years ago, the universe originated with such a force that the farthest galaxies are hurling away from

us at nearly the speed of light. The beginning universe might have rapidly ballooned outward for an instant, a phenomenon called inflation, and then settled down to a steady growth. The expanding fireball required perhaps 300,000 years to cool sufficiently to allow the basic units of matter to clump together to form billions of galaxies each containing billions of stars (Fig. 1). The first galaxies evolved when the universe was about 1 billion years old and only about a tenth of its present size.

Our Milky Way is an elliptical galaxy, with five spiral arms peeling off a central bulge. New stars originate in dense regions of interstellar gas and dust called giant molecular clouds. Several times a century, a giant star more than 100 times larger than the Sun explodes, producing a supernova a billion times brighter than an ordinary star. When a star reaches the supernova stage, after a very hot existence spanning several hundred million years, the nuclear reactions in its core become highly explosive. While the star sheds its outer covering, the core compresses to an extremely dense, hot body called a neutron star, similar to condensing Earth down to about the size of a golf ball.

The expanding stellar matter from the supernova forms a nebula composed mostly of hydrogen and helium along with particulate matter that comprises all the other known elements. About a million years later, the solar nebula collapses into a star. This process begins when shock waves from nearby supernovas compress portions of the nebula, causing the nebular matter to collapse upon itself by gravitational forces into a protostar. As the solar nebula collapses, it rotates faster and faster, and spiral arms peel off the rapidly spinning nebula to form a protoplanetary disk. Meanwhile, the compressional heat initiates a thermonuclear reaction in the core, and a star is born.

A new star forms in the Milky Way galaxy every few years or so. About 4.6 billion years ago, the Sun, an ordinary main-sequence star, ignited in one of the dusty spiral arms of the galaxy about 30,000 light-years from the center. Single, medium-size stars such as the Sun are a rarity. Due to their unique evolution, these stars appear to be the only ones with planets. Thus, of the myriad of stars overhead, only a handful might possess a system of orbiting planets, and fewer yet might contain life.

When the Sun first ignited, the strong solar wind blew away the lighter components of the solar nebula and deposited them into the outer regions of the solar system. The remaining matter in the inner solar system comprised mostly stony and metallic fragments, ranging in size from fine sand grains to huge boulders. In the outer solar system, where temperatures were much colder, rocky material along with solid chunks of water ice, frozen carbon dioxide, and crystalline methane and ammonia condensed. The outer planets are believed to possess rocky cores about the size of Earth, surrounded by a mantle composed of water ice and frozen methane, and a thick atmosphere, mostly composed of hydrogen and helium.

The Sun was extremely unstable during its first billion years of existence. The solar output was only about 70 percent of its present intensity. The Sun provided only as much warmth on Earth as it presently does on Mars. The early Sun rapidly spun on its axis, completing a single rotation in just a few days as compared with 27 days today. Periodically, the Sun expanded up to a third larger than its current size, and enormous solar flares leaped millions of

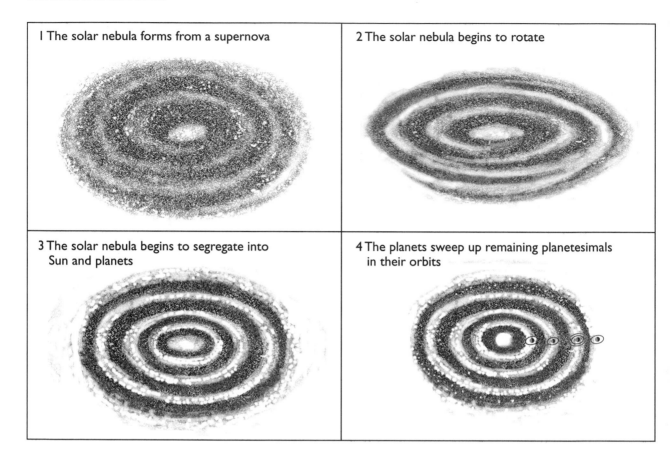

1 The solar nebula forms from a supernova

2 The solar nebula begins to rotate

3 The solar nebula begins to segregate into Sun and planets

4 The planets sweep up remaining planetesimals in their orbits

Figure 2 *The formation of the solar system from planetesimals in the solar nebula.*

miles into space. This intense activity made the Sun radiate more heat, allowing it to cool and return to its normal size.

During the Sun's early developmental stages, it was ringed by a protoplanetary disk composed of several bands of coarse particles, called planetesimals that accreted from grains of dust cast off by a supernova. Some 100 trillion planetesimals orbited the Sun during the solar system's early stages of development (Fig. 2). As they continued to grow, the small rocky chunks swung around the infant Sun in highly elliptical orbits along the same plane called the ecliptic.

The constant collisions among planetesimals built larger bodies, some of which grew to more than 50 miles wide. However, most of the planetary mass still resided in the small planetesimals. The presence of a large amount of gas in the solar nebula slowed the planetesimals, enabling them to coalesce into planets. The planetesimals in orbit between Mars and Jupiter were unable to combine into a planet due to Jupiter's strong gravitational attraction and instead formed a belt of asteroids, many of which were several hundreds miles wide.

An incredible amount of water, one of the simplest of molecules, resides in the solar system. As the Sun emerged from gas and dust, tiny bits of ice and rock debris began to gather in a frigid, flattened disk of planetesimals surrounding the infant star. The temperatures in some parts of the disk might have been warm enough for liquid water to exist on the first solid bodies in the solar system. In addition, water vapor in the primordial atmospheres of the inner terrestrial planets might have eroded away by planetesimal bombardment and blown beyond Mars by the strong solar wind of the infant Sun. Once planted in the far reaches of the solar system, the water coalesced to create icy bodies that streak by the Sun as comets.

The solar system is quite large, consisting of nine known planets and their moons (Fig. 3), although controversy remains whether Pluto is actually a planet or some other type of body. The image of the original solar disk can be traced by observing the motions of the planets. All of them revolve around the Sun in the same direction it rotates, and all but one, Pluto, orbit within 3 degrees of the ecliptic. Because Pluto's orbit inclines 17 degrees to the ecliptic, it might be a captured planet or possibly a moon of Uranus knocked out of orbit by collision with another body or a comet from the Kuiper belt.

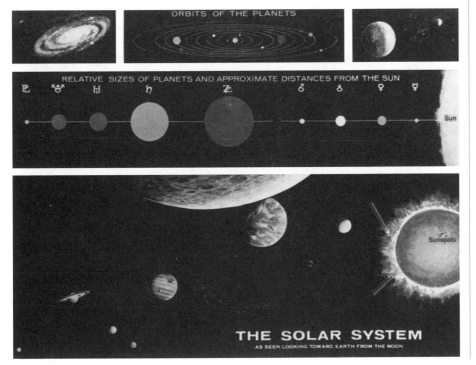

Figure 3 *The structure of the solar system.*

(Photo courtesy NASA)

Some 7 billion miles from the Sun lies the heliopause, which marks the boundary between the Sun's domain and interstellar space. About 20 billion miles from the Sun is a region of gas and dust, possibly remnants of the original solar nebula. A ring of comets that make up the Kuiper belt lies on the ecliptic in this region. Several trillion miles from the sun is a shell of comets called the Oort cloud that formed from the leftover gas and ice of the original solar nebula.

THE PROTOEARTH

Around 4.6 billion years ago, the primal Earth emerged from a spinning, turbulent cloud of gas, dust, and planetoids surrounding an infant star. During the next 700 million years, the cloud settled into a more tranquil solar system, and the Sun's third planet began to take shape. The molten Earth continued to grow by accumulating planetesimals, most of which were hot with temperatures exceeding 1,000 degrees Celsius.

Due to drag forces created by leftover gases in interplanetary space, Earth's orbit began to decay. The formative planet slowly spiraled inward toward the Sun, sweeping up additional planetesimals along the way like a cosmic vacuum cleaner. Eventually, Earth's path around the Sun was swept clean of interplanetary material, leaving a gap in the disk of planetesimals, and its orbit stabilized near its present position.

The core and mantle (Fig. 4) segregated possibly within the first 100 million years during a time when Earth was in a molten state caused by radioactive isotopes and impact friction from planetesimals. The presence of magnetic rocks as old as 3.5 billion years suggests that Earth had a molten outer core and a solid inner core comparable to their present sizes at an early age. The core attracted siderophilic, or iron-loving, materials such as gold, platinum, and certain other elements originating from meteorite bombardments during the planet's early formation.

Earth's interior was hotter, less viscous, and more vigorous, with a highly active convective flow. Heavy turbulence in the mantle with a heat flow three times greater than that today produced violent agitation on Earth's surface. This turmoil created a sea of molten and semimolten rock broken up by giant fissures, from which fountains of lava spewed skyward.

During the first half-billion years, Earth's surface was scorching hot. Heat of compression from the primordial atmosphere with pressures a hundred times greater than that today resulted in surface temperatures hot enough to melt rocks. When the Sun ignited, the strong solar wind blew away the lighter components of Earth's atmosphere. Meanwhile, a massive bombardment of meteorites blasted the remaining gases into space, leaving the planet in a near vacuum much like the Moon today.

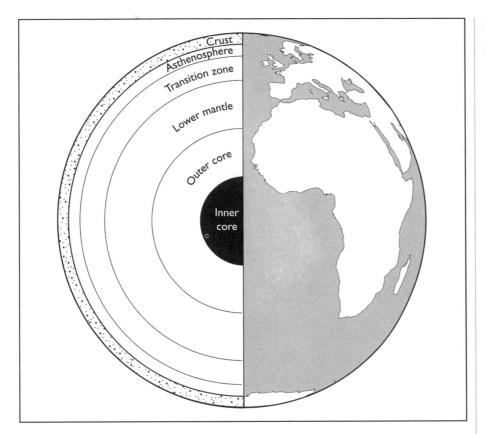

Figure 4 *The structure of Earth, showing the core, mantle, and crust.*

In the absence of an atmosphere to hold in the internally generated heat, the surface rapidly cooled, forming a thin basaltic crust similar to that of Venus. Indeed, the Moon and the inner planets offer clues to Earth's early history. Among the features common to the terrestrial planets was their ability to produce voluminous amounts of basaltic lavas. Earth's original crust has long since disappeared, remixed into the interior by the impact of giant meteorites that were leftovers from the creation of the solar system.

The formative Earth was subjected to extensive volcanism and intense meteorite bombardment that repeatedly destroyed the crust. A massive meteorite shower, consisting of thousands of 50-mile-wide impactors, bombarded Earth and the Moon around 3.9 billion years ago. The other inner planets and the moons of the outer planets show dense pockmarks from this invasion (Fig. 5). The meteorite bombardment melted large portions of Earth's crust, nearly half of which contained large impact basins up to 10 miles deep.

As the meteorites plunged into the planet's thin basaltic crust, they gouged out huge quantities of partially solidified and molten rock. The scars

Figure 5 *Saturn's heavily cratered moon Mimas from* Voyager I *in November 1980.*

(Photo courtesy NASA)

in the crust quickly healed as batches of fresh magma bled through giant fissures and poured onto the surface, creating a magma ocean. The continued destruction of the crust by heavy volcanic and meteoritic activity explains why the first 700 million years of Earth history, referred to as the Hadean or Azoic eon, are missing from the geologic record.

The original crust was quite distinct from modern continental crust, which first appeared about 4 billion years ago and now represents less than 0.5 percent of the planet's total volume. During this time, Earth spun wildly on its axis, completing a single rotation every 14 hours, thus maintaining high temperatures throughout the planet. Present-day plate tectonics (the interaction of crustal plates) could not have operated under such hot conditions, which produced more vertical bubbling than horizontal sliding. Therefore, modern-style plate tectonic processes were probably not fully functional until about 3 billion years ago, when the formation of the crust was nearly complete.

Much information about the early crust is provided by some of the most ancient rocks that survived intact. They formed deep within the crust a few hundred million years after the formation of the planet and now outcrop at the surface. Zircon crystals (Fig. 6) found in granite are enormously resistant

Figure 6 Zircons from the rare-earth zone, Jasper Cuts area, Gilpin County, Colorado.

(Photo by E. J. Young, courtesy USGS)

and tell of the earliest history of Earth, when the crust first formed some 4.2 billion years ago.

Among the oldest rocks are those of the 4–billion–year–old Acasta gneiss, a metamorphosed granite in the Northwest Territories of Canada. Its existence suggests that the formation of the crust was well underway by this time. The discovery is used as evidence that at least small patches of continental crust existed on Earth's surface at an early date. Earth apparently took less than half its history to form an equivalent volume of continental rock as it has today.

THE MOON

A popular contention for the creation of the Moon supposes a collision between Earth and a large celestial body. According to this theory, soon after Earth's formation, an asteroid about the size of Mars was knocked out of the asteroid belt either by Jupiter's strong gravitational attraction or by a collision with a wayward comet. On its way toward the inner solar system, the asteroid glanced off Earth (Fig. 7). The tangential collision over a period of half an hour created a powerful explosion equivalent to the detonation of an amount of dynamite equal to the mass of the asteroid. The collision tore a huge gash in Earth. A large portion of its molten interior along with much of the rocky mantle of the impactor spewed into orbit, forming a ring of debris around the planet called a protolunar disk, similar to the rings of Saturn.

The force of the impact might have knocked Earth over, tilting its rotational axis about 25 degrees. Similar collisions involving the other planets, especially Uranus, which orbits on its side like a rolling ball, might explain their various degrees of tilt and elliptical orbits. The glancing blow might also have increased Earth's angular momentum (rotational energy) and melted the planet throughout, forming a red-hot orb in orbit around the Sun. As a result, Earth would have spun wildly on its axis, completing a day in only two hours. The present angular momentum suggests that other methods of lunar formation such as fission, capture, or assembly at the same time as Earth were unlikely.

Clinching evidence for the collision theory of lunar origin was obtained by analyzing moon rocks (Fig. 8) brought back during the Apollo missions of the late 1960s and early 1970s. The rocks appear to be similar in composition to Earth's upper mantle and range in age from 4.5 to 3.2 billion years old. Since no rocks were found younger than this date, the Moon probably ceased volcanic activity, and the interior began to cool and solidify.

The new satellite continued growing as it swept up debris in orbit around Earth. In addition, huge rock fragments orbited the Moon and crashed onto its surface. The massive meteorite shower that bombarded Earth equally pounded the Moon. Many large asteroids struck the lunar surface and broke

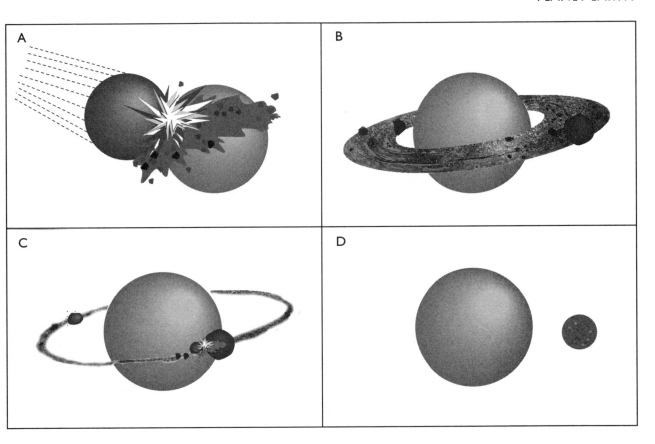

through the thin crust. Great floods of dark basaltic lava spilled onto the surface, giving the Moon a landscape of giant craters and flat lava plains (Fig. 9) called *maria* from the Latin word for "seas."

The Moon became gravitationally locked onto its mother planet, rotating at the same rate as its orbital period, causing one side always to face Earth. Many moons around other planets share similar characteristics with the Earth's moon, suggesting they were created in much the same manner. Since Earth's sister planet Venus formed under similar circumstances and is so much like our planet, the absence of a Venusian moon is quite curious. It might have crashed into its mother planet or escaped into orbit around the Sun. Perhaps Mercury, which is about the same size as the Earth's moon, was once a moon of Venus.

The newly formed moon circled just 14,000 miles above Earth, racing around the planet every two hours. It orbited so close to Earth it filled much of the sky. The Moon's nearness also explains why the length of day was so much shorter. The early Earth spun faster on its axis than it does today. As the planet slowed its rotation due to drag forces caused by the tides (making days

Figure 7 *The giant impact hypothesis of lunar formation envisions a Mars-sized planetesimal (A) colliding with the protoearth (B) resulting in a gigantic explosion and the jetting outward of both planetesimal and protoearth material into orbit around the planet. (C) A protomoon begins to form from a prelunar disk, and matter accretes to form the Moon (D).*

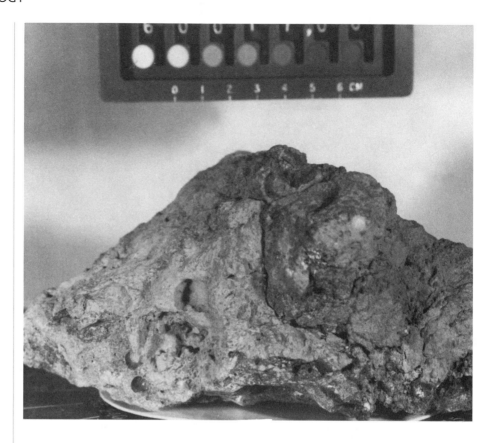

longer), it transferred some of its angular momentum to the Moon, flinging it outward into a widening orbit. Eventually, the Moon's orbit gradually widened to 240,000 miles away. Even today, the Moon is still receding from the Earth about an inch or so each year.

The presence of a rather large moon, the biggest in relation to its mother planet, the two forming a twin planetary system, might have had a major influence on the initiation of life. The unique properties of the Earth-Moon system raised tides in the ocean. Cycles of wetting and drying in tidal pools might have helped Earth acquire life much earlier than previously thought possible.

The Moon might also have been responsible for the relatively stable climate, making Earth hospitable to life by stabilizing the tilt of the planet's rotational axis, which marks the seasons. Without the Moon, life on Earth would likely face the same type of wild fluctuations in climate that Mars has apparently experienced through the eons. If its spin axis were no longer maintained by the Moon, in only a few million years Earth would have drastically altered its tilt sufficiently to make the polar regions warmer than the Tropics.

THE ATMOSPHERE

For the first half billion years, while Earth was spinning wildly on its axis and surface rocks were scorching hot, the planet lacked an atmosphere. It was in a near vacuum much like the Moon is today. Soon after the meteorite bombardment began about 4.2 billion years ago, Earth acquired a primordial atmosphere composed of carbon dioxide, nitrogen, water vapor, and other gases spewed out of a profusion of volcanoes along with gases such as ammonia and methane delivered by a fussilade of comets. The atmosphere was so saturated with water vapor that atmospheric pressure was nearly 100 times greater than it is today.

Much of the water vapor and gases came from outer space. Some meteorites that visited Earth were stony composed of rock and metal, others were icy composed of frozen gases and water ice, and many contained carbon as

Figure 9 View of a full Moon taken from Apollo 11 *spacecraft, showing numerous craters and expansive lava plains.*

(Photo courtesy NASA)

though millions of tons of coal rained down from the heavens. Perhaps these carbon-rich meteorites bore the seeds of life, which might have existed in the universe eons before Earth came into being. Comets, composed of rock debris and ice, also plunged into Earth, releasing large quantities of water vapor and gas. These cosmic gases were mostly carbon dioxide, ammonia, and methane.

Most of the water vapor and gases originated within Earth itself. Magma contains large quantities of volatiles, mostly water and carbon dioxide, which made it more fluid. Tremendous pressures deep inside Earth kept the volatiles within the magma. However, when the magma rose to the surface, the drop in pressure released the trapped water and gases, often explosively. The early volcanoes erupted violently because Earth's interior was much hotter and the magma contained higher amounts of volatiles.

The early atmosphere contained up to 1,000 times the current level of carbon dioxide. This was fortunate because the Sun's output was only about three-quarters of its present value, and a strong greenhouse effect kept Earth from freezing solid. The planet also retained its warmth by a high rotation rate and by the absence of continents to block the flow of ocean currents.

Oxygen originated directly by volcanic outgassing and meteorite degassing. It was also produced indirectly by the breakdown of water vapor and carbon dioxide by strong ultraviolet radiation from the Sun. All oxygen generated in this manner quickly bonded to metals in the crust, much like the rusting of iron. Oxygen also recombined with hydrogen and carbon monoxide to reconstitute water vapor and carbon dioxide. A small amount of oxygen might have existed in the upper atmosphere, where it provided a thin ozone screen. This would have reduced the breakdown of water molecules by strong ultraviolet rays from the Sun and prevented the loss of the entire ocean, a fate that might have visited Venus eons ago (Fig. 10).

Nitrogen, which comprises 79 percent of the present atmosphere, originated from volcanic eruptions and from the breakdown of ammonia, a molecule with one nitrogen atom and three hydrogen atoms. Ammonia was a major constituent of the primordial atmosphere. Unlike most other gases, which have been replaced or recycled, Earth retains much of its original nitrogen. This is because nitrogen readily transforms into nitrate, which easily dissolves in the ocean, where denitrifying bacteria return the nitrate-nitrogen to its gaseous state. Without life, Earth would have long ago lost its nitrogen and possess only a fraction of its present atmospheric pressure.

THE OCEAN

During the formation of the atmosphere, Earth's surface was constantly in chaos. Winds blew with a tornadic force. Fierce dust storms on the dry sur-

Figure 10 *A radar image collection of Venus obtained by the Magellan mission in 1994.*

(Photo courtesy NASA)

face blanketed the entire planet with suspended sediment much like the Martian dust storms of today (Fig. 11). Huge lightning bolts flashed across the sky. The thunder was earth-shattering as one gigantic shock wave after another reverberated throughout the land. Volcanoes erupted in one giant outburst after another. The sky lit up from the pyrotechnics created by the white-hot sparks of ash and the glow of constantly flowing red-hot lava. The restless Earth was rent apart as massive quakes cracked open the thin crust. Huge batches of magma flowed through the fissures and flooded the surface with voluminous amounts of lava, forming flat, featureless basalt plains.

The intense volcanism lofted millions of tons of volcanic debris into the atmosphere, where it remained suspended for long periods. Ash and dust particles scattered sunlight and gave the sky an eerie red glow like that on Mars following an intense dust storm. The dust also cooled Earth and provided particulate matter, upon which water vapor could coalesce. When temperatures in the upper atmospheric lowered, water vapor condensed into clouds. The clouds were so thick and heavy they almost completely blocked out the Sun. As a result, the surface was in near darkness, dropping temperatures even further.

As the atmosphere continued to cool, huge raindrops fell from the sky, and Earth received deluge after deluge. Raging floods cascaded down steep

Figure 11 *The landscape of Mars from* Viking I, *showing boulders surrounded by windblown sediment.*

(Photo courtesy NASA)

mountain slopes and the sides of large meteorite craters, gouging out deep canyons in the rocky plain. When the rains ceased and the skies finally cleared, Earth emerged as a giant blue orb, covered by a global ocean nearly 2 miles deep and dotted with numerous volcanic islands.

Ancient marine sediments found in the metamorphosed rocks of the Isua Formation in southwestern Greenland support this scenario for the creation of the ocean. The rocks originated in volcanic island arcs and therefore lend credence to the idea that some form of plate tectonics operated early in the history of Earth. They are among the oldest rocks, dating to about 3.8 billion years ago, and indicate that the planet had surface water by this time.

During the intervening years between the end of the great meteorite bombardment and the formation of the first sedimentary rocks, vast quantities of water flooded Earth's surface. Seawater probably began salty due to the abundance of chlorine and sodium provided by volcanoes. However, the ocean did not reach its present concentration of salts until about 500 million years ago. The warm seas were heated from above by the Sun and from below by active volcanoes on the ocean floor, which continually supplied seawater with the elements of life.

THE EMERGENCE OF LIFE

Life arose on this planet during a period of crustal formation and volcanic outgassing of an atmosphere and ocean (Table 1). It was also a time of heavy meteorite bombardment. Rocky asteroids and icy comets constantly showered the early Earth, possibly providing the main source of the planet's water. Inter-planetary space was littered with debris that pounded the newborn planets. Some of this space junk might have supplied organic compounds, from which life could evolve.

Earlier theories on the creation of life relied on the so-called primordial soup hypothesis. To test the theory, a spark discharge chamber (a device first designed in the 1950s to replicate the early atmosphere and ocean) was built to represent prebiotic conditions on the early Earth (Fig. 12). As a result, all the precursors of life (the elements within the chamber: methane, ammonia, and hydrogen) came together. By countless combinations and permutations, an organic molecule evolved that could replicate itself. However, the time

---

TABLE 1   EVOLUTION OF THE BIOSPHERE

	Billions of Years Ago	Percent Oxygen	Biologic Effects	Event Results
Full oxygen conditions	0.4	100	Fish, land plants, and animals	Approach present biologic environs
Appearance of shell-covered animals	0.6	10	Cambrian fauna	Burrowing habitats
Metazoans appear	0.7	7	Ediacaran fauna	First metazoan fossils and tracks
Eukaryotic cells appear	1.4	>1	Larger cells with a nucleus	Redbeds, multi-cellular organisms
Blue-green algae	2.0	1	Algal filaments	Oxygen metabolism
Algal precursors	2.8	<1	Stromatolite mounds	Initial photo-synthesis
Origin of life	4.0	0	Light carbon	Evolution of the biosphere

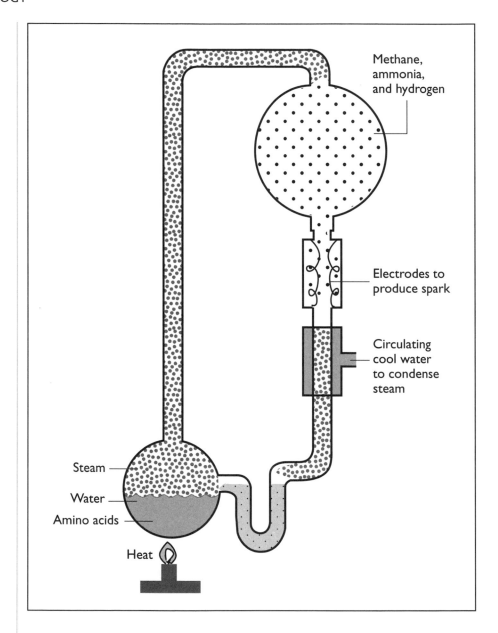

Figure 12 *A spark discharge chamber represents prebiotic conditions on the early Earth.*

Methane, ammonia, and hydrogen

Electrodes to produce spark

Circulating cool water to condense steam

Steam

Water

Amino acids

Heat

required for such a random event would involve billions of years. Moreover, evidence gathered from ancient rocks on Earth, the Moon, and meteorites suggests that the amount of ammonia and methane assumed to be in the primordial atmosphere was not nearly as abundant as originally thought.

Biophysicists, who study the mechanics of life, have discovered intriguing evidence in the interior of the 4.5-billion-year-old Murchison meteorite, named for a site in Western Australia where it fell to Earth in 1969. The meteorite held lipidlike organic compounds able to self-assemble into cell-like membranes—an essential requirement for the first living cells. The meteorite, a carbonaceous chondrite, is believed to have broken off an asteroid that formed about the same time and from similar materials as Earth. The organic chemicals provided the first unambiguous evidence of extraterrestrial amino acids. The material in the meteorite thus contains many essential components necessary for creating life.

Earth is still pelted by meteorites that contain amino acids, the precursors of proteins. The early meteorite impacts would also most likely have made conditions very difficult for proteins to organize into living cells. The first cells might have been repeatedly exterminated, forcing life to originate over and over. Whenever primitive organic molecules attempted to arrange themselves into living matter, frequent impacts blasted them apart before they had a chance to reproduce.

Some large impactors might have generated enough heat to evaporate most of the ocean many times. The vaporized ocean would have raised surface pressures more than a hundred times greater than the present atmosphere, and the resulting high temperatures would have sterilized the entire planet. Several thousand years would elapse before the steam condensed into rain and the ocean basins refilled again, only to await the next ocean-evaporating impact. Such harsh conditions could have set back the emergence of life hundreds of millions of years.

Perhaps the only safe place for life to evolve was 3 to 4 miles down on the deep ocean floor, where a high density of hydrothermal vents existed. Hydrothermal vents are like geysers on the bottom of the sea (Fig. 13) that expel mineral-laden hot water heated by shallow magma chambers resting just beneath the ocean floor. The vents might have created an environment capable of generating an immense number of organic reactions and could have provided the ingredients and energy needed to create the planet's first life. They would also have given evolving life-forms all the essential nutrients needed to sustain themselves. Indeed, such an environment exists today and is home to some of the strangest creatures found on Earth. In this environment, life could have originated as early as 4.2 billion years ago.

From the very beginning, life had many common characteristics. No matter how varied organisms are today, from the simplest bacteria to ourselves, its central molecular machinery is exactly the same. Every cell of every species is constructed from the same set of 20 amino acids. All life-forms use the same energy transfer mechanism for growth. All strands of DNA are left-handed double helixes, and the operation of the genetic code in protein synthesis is the same for all living things.

With so much similarity, all life must have sprung from a common ancestor. Any alien forms, of which no descendants exist today, became extinct

Figure 13 *An active hydrothermal vent and sulfide mineral deposits at the East Pacific Rise.*

(Photo courtesy USGS)

early in the history of life. Furthermore, no new life-forms are being created today either because the present chemical environment is not conducive to the formation of life or living organisms prey upon the newly created organisms before they have any chance of evolving.

Since life appeared within the first half-billion years of Earth's existence, it must have evolved into complex organisms from simple materials rather quickly. Primitive bacteria, which descended from the earliest known form of life, remain by far the most abundant living beings. Evidence that life began early in Earth's history when the planet was still quite hot exists today as thermophilic (heat-loving) bacteria, found in thermal springs and other hot-water environments throughout the world (Fig. 14) as well as deep underground or far below the ocean floor.

The existence of these organisms is evidence that thermophiles were the common ancestors of all life. The early conditions on Earth would have been ripe for the evolution of thermophilic organisms. Most of these have a sulfur-based energy metabolism, and sulfur compounds would have been plentiful on the hot, volcanically active planet.

Fortunately for the early Earth, it had an abundance of sulfur, which spewed out of a profusion of volcanoes. As long as surface temperatures remained hot, ring molecules of sulfur atoms in the atmosphere would block out solar ultraviolet radiation. Otherwise, the first living cells would have sizzled in the deadly rays of the Sun. However, an ultraviolet shield might not have been necessary in the primordial atmosphere, because some primitive bacteria appear to tolerate high levels of ultraviolet radiation.

Evidence that life began quite early in Earth history when the planet was steaming hot exists today as archaebacteria, or simply archaea. They range more widely than previously believed, and many parts of the ocean are teeming with them. A third of the biomass of picoplankton (the tiniest plankton) in Antarctic waters were archaea. Such abundance could mean that archaea play an important role in the global ecology and might have a major influence on the chemistry of the ocean.

The first living organisms were extremely small noncellular blobs of protoplasm. The self-duplicating organisms fed on a rich broth of organic molecules generated in the primordial sea. Such a nutritional abundance set off a rapid chain reaction, resulting in phenomenal growth. The organisms drifted freely in the ocean currents and dispersed to all parts of the world. Although

Figure 14 *Hot carbonated springwater undercuts bedded travertine deposits at Yellowstone National Park, Wyoming.*

(Photo by K. E. Barger, courtesy USGS)

the first simple organisms appeared to have arrived soon after conditions on Earth became favorable, almost another billion years passed before life even remotely resembled anything living today.

After learning how Earth evolved, the next chapter will examine the Archean eon and the earliest life-forms along with the earliest rocks.

2

ARCHEAN ALGAE

THE AGE OF EARLY LIFE

This chapter examines the earliest history of Earth, its evolving life-forms, and the formation of continents. The first 4 billion years, or about nine-tenths of geologic time (Table 2), are referred to as the Precambrian era, the longest and least understood period of Earth history. The Precambrian began with only simple creatures living in the sea. It ended with an explosion of new, highly specialized species, which set the stage for more modern life-forms to follow (Fig. 15)

The Precambrian is divided nearly equally into the Archean and Proterozoic eons. The boundary is somewhat arbitrary and reflects major differences between the types of rocks formed during the two periods. Archean rocks are products of rapid crustal formation. Proterozoic rocks are more representative of relatively stable modern geology.

The Archean, from 4.6 to 2.5 billion years ago, covers a time when Earth was in a great turmoil and subjected to extensive volcanism and intense meteorite bombardment. The high internal heat of the newborn planet kept the surface well agitated, destroying any semblance of a crust. This is why the first several hundred million years are absent from the geologic record. During this

TABLE 2   THE GEOLOGIC TIME SCALE

Era	Period	Epoch	Age (Millions of Years)	First Life-Forms	Geology
		Holocene	0.01		
	Quaternary				
		Pleistocene	3	Humans	Ice age
Cenozoic		Pliocene	11	Mastodons	Cascades
		Neogene			
		Miocene	26	Saber-toothed tigers	Alps
	Tertiary	Oligocene	37		
		Paleogene			
		Eocene	54	Whales	
		Paleocene	65	Horses, Alligators	Rockies
	Cretaceous		135		
				Birds	Sierra Nevada
Mesozoic	Jurassic		210	Mammals	Atlantic
				Dinosaurs	
	Triassic		250		
	Permian		280	Reptiles	Appalachians
	Pennsylvanian		310		Ice age
				Trees	
	Carboniferous				
Paleozoic	Mississippian		345	Amphibians	Pangaea
				Insects	
	Devonian		400	Sharks	
	Silurian		435	Land plants	Laursia
	Ordovician		500	Fish	
	Cambrian		570	Sea plants	Gondwana
				Shelled animals	
			700	Invertebrates	
Proterozoic			2500	Metazoans	
			3500	Earliest life	
Archean			4000		Oldest rocks
			4600		Meteorites

interval, the planet experienced a restlessness that might have been a major factor in the emergence of life so early in the history of Earth.

THE AGE OF ALGAE

Life in the Archean consisted mostly of bacteria, unicellular algae, and clusters of algae called stromatolites (Fig. 16), from the Greek word *stroma* meaning "stony carpet." The oldest evidence of life includes microfossils and stromatolites. Microfossils are the remains of ancient microorganisms. Stromatolites are layered structures formed by the accretion of fine sediment grains by matted colonies of cyanobacteria or primitive blue-green algae living in shallow seas. In addition, mats of microbes as much as one-third inch

Figure 15 *The geologic time spiral reveals the history of the Earth.*

(Photo courtesy USGS)

25

Figure 16 Stromatolites west of Logan Pass, Glacier National Park, Montana.

(Photo by R. Rezak, courtesy USGS)

thick grew on the surface of a clay-rich soil, becoming the first life on land some 2.6 billion years ago.

The stromatolite colonies formed from layers of cells topped by photosynthetic organisms that multiplied using sunlight and supplied the lower layers with nutrients. Stromatolites are only indirect evidence of early life, however. They are not the remains of the microorganisms themselves but only the sedimentary structures they built. Early stromatolite fossils exist in 3.5-billion-year-old sedimentary rocks of the Towers Formation of the Warrawoona Group in North Pole, Western Australia. The region was once a tidal inlet, overshadowed by tall volcanoes that erupted ash and lava, which flowed into a shallow sea. Thunderclouds hovered over the peaks, and lightning darted back and forth. Furious winds whipped up high waves that pounded the basaltic cliffs of the coastline.

Farther inland, hummocks (rounded piles) of black basalt flows, still steamy from their latest eruptions, dominated the landscape. The rotten egg stench of sulfur was pervasive. Frequent downpours fed tidal streams that meandered onto a flat expanse of glistening gray mud before reaching the sea. Elsewhere, scattered shallow pools containing highly saline water periodically evaporated, leaving behind a variety of salts. Often, a flood tide washed across the mudflat, shifting the sediments and replenishing the brine pools.

Although the Archean spans almost half of Earth's history, its rocks represent less than 20 percent of the total area exposed at the surface. Furthermore, all known Precambrian rocks have suffered some heating episode and metamorphism. Unlike most ancient rocks in the 3.5- to 3.8-billion-year range throughout the world, only a few such as those of the North Pole sequence have a history of low metamorphic temperatures. Therefore, rocks of this region have remained relatively cool throughout geologic history.

Rocks subjected to the intense heat of Earth's interior have lost all traces of fossilized life. Even in mildly metamorphosed rocks, the existence of microfossils, which are the preserved cell walls of unicellular microorganism, is often difficult to prove. Most of these apparent fossils are simple spheres with few surface features. They are composed of inorganic carbon compounds squeezed into spheroids by the growth of mineral grains deposited around them. However, some spheres were linked in pairs or in chains, and others were in groups of four and unlikely created simply by inorganic processes.

Associated with the North Pole rocks were cherts, extremely hard siliceous rocks containing microfilaments, which are small, threadlike structures of possible bacterial origin. Similar cherts with microfossils of filamentous (threadlike) bacteria are found at eastern Transvaal, South Africa, and date between 3.2 and 3.3 billion years old. They also exist in 2-billion-year-old chert from the Gunflint Iron Formation on the north shore of Lake Superior in North America. Most Precambrian cherts appears to be chemical sediments precipitated from silica-rich water in deep oceans. The abundance of chert in the Archean might serve as evidence that most of the crust was deeply submerged. However, cherts in the North Pole region appear to have a shallow-water origin.

Chert-forming silica leached out of volcanic rocks that erupted into shallow seas. The silica-rich water circulated through porous sediments, dissolving the original minerals and precipitating silica in their place. Microorganisms buried in the sediments were thus encased in one of nature's hardest substances, enabling the microfossils to survive the rigors of time. Modern seawater is deficient in silica because organisms such as sponges and diatoms (Fig. 17) extract it to build their skeletons. Generally, only the silica spicules that make up the skeletons of sponges remain as fossils. Diatoms exhibit beautiful glasslike silica laceworks within their cell walls. Massive deposits of diatomaceous earth, also called diatomite, composed of diatom cell walls are a tribute to the great success of these organisms following the Precambrian.

The North Pole stromatolites are distinctly layered accumulations of calcium carbonate with a rounded, cabbagelike appearance. The size and shape of the Archean-age microfossils and the form of the stromatolites suggest these microorganisms were either oxygen-releasing or sulfur-oxidizing photosynthetic species dependent on sunlight for their growth. Sulfur-metabolizing bac-

Figure 17 *Late Miocene marine diatoms from the Kilgore area, Cherry County, Nebraska.*

(Photo by G. W. Andrews, courtesy USGS)

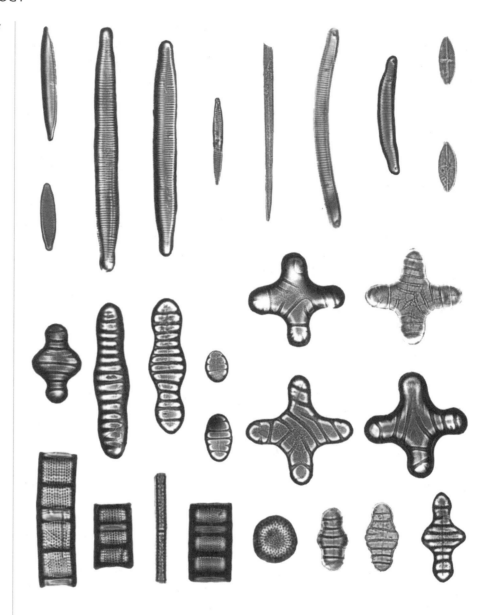

teria might have arisen in the earliest stages of biologic evolution, probably about 3.5 billion years ago. Iron-metabolizing bacteria might have been responsible for the great banded iron formations laid down some 2 billion years ago.

Modern stromatolites reside in the intertidal zones above the low-tide mark. Their length reflects the height of the tides, controlled mostly by the gravitational pull of the Moon. The oldest stromatolite colonies of the North Pole region grew to great heights, with some attaining lengths of more than

30 feet. The evidence suggests that at an early age, the Moon orbited much closer to Earth, and its strong gravitational attraction at this range raised tremendous tides that flooded coastal areas long distances inland.

Living stromatolites are similar to those of ancient times. They comprise concentrically layered mounds of calcium carbonate built by algae or bacteria, which cement sediment grains together by secreting a jellylike ooze. Older structures at the Australian site are classified either as stromatolite fossils or as layered inorganic sedimentary structures. However, microscopic filaments radiating outward from a central point resembling filamentous bacteria also exist in the fossils, which suggests bacteria built the stromatolites.

Bacteria are life's greatest success story. They occupy a wider domain of environments and span a broader range than any other group of organisms. They are extremely adaptable, indestructible, astoundingly diverse, and absolutely necessary for the existence of other life-forms. The bacterial mode of life has been stable from the very beginning of the fossil record. Long after all other species have gone extinct as Earth nears its end, bacteria will remain as they were at the beginning—the only life on Earth.

THE PROTOZOANS

Protozoans are primitive animals that have survived throughout three-quarters of Earth history. They are often classified into the kingdom Protistae, which includes all single-celled plants and animals with a nucleus. In the obscured past, few distinctions existed between early plants and animals. They shared many similar characteristics. The protozoans are also classified into the animal kingdom, and indeed the term literally means "beginning animals."

For most of the Archean, the only life-forms were simple organisms with primitive cells called prokaryotes, derived from the Greek word *karyon* meaning "nutshell." They lacked a distinct nucleus and lived under anaerobic (without oxygen) conditions. They depended mainly on outside sources of nutrients, typically a rich supply of organic molecules continuously being created in the sea around them. Most organisms had a primitive form of metabolism called fermentation that converted nutrients into energy. This was an inefficient form of metabolism, releasing energy when enzymes broke down simple sugars such as glucose into smaller molecules.

A more advanced organism comprised a committee of simpler organelles about the size of prokaryotes that the organism incorporated into its cell in a symbiotic (helpful) relationship, creating a new type of life-form called a eukaryote. It was equipped with a nucleus that organized genetic material and includes all plants, animals, fungi, and algae. The divergence of prokaryotes and eukaryotes probably began before 3 billion years ago. However, the development of the eukaryotic cell possibly took as long as 1 billion years before it resembled anything living today. Eukaryotic cells are typically 10,000 times larger than prokaryotes.

During cell division, DNA in the nucleus and in the organelles replicated, with half the genes remaining with the parent and the other half passed on to the daughter cell. This process, called mitosis, increased the likelihood of genetic variation. It greatly accelerated the rate of evolution as organisms encountered new environments and could adapt. The extraordinary variety of plant and animal life that has arisen on this planet over the last 600 million years is due exclusively to the introduction of the eukaryotic cell and its huge potential for genetic diversity.

Early single-cell animals called protists were the progenitors of all other animal species. They were the first group of organisms to evolve a nucleus and shared many characteristics with plants. The cells contained elongated structures of mitochondria, which were bacterialike bodies that produce energy by oxidation. They also contained chloroplasts, which were packets of chlorophyll that provide energy by photosynthesis.

Many protozoans secreted a tiny shell composed of calcium carbonate. When the animals died, their shells sank to the bottom of the ocean like a constant rain. Over time, they built up impressive formations of limestone (Fig. 18). The shifting of these sediments by storms and undersea currents buried dead marine organisms that were not eaten by scavengers. A calcite ooze was then formed, which eventually hardened into limestone, preserving trapped species for all time.

Some varieties of protists formed large colonies, whereas most lived independently. The entire body was composed of a single cell containing living protoplasm enclosed within a membrane. Present-day single-celled organisms have not changed significantly from ancient fossils. However, most archaic forms were soft bodied and did not fossilize well. They obtained energy by ingesting food particles or by photosynthesis and reproduced by using unicellular gametes that do not form embryos.

The major groups of protists include algae, diatoms, dinoflagellates, fusulinids, radiolarians, and foraminiferans (Fig. 19). The foraminiferans were microscopic protozoans whose skeletons composed of calcium carbonate preserved much of the record of the behaviour of the ocean and climate. Most lived on the bottom of shallow seas. A few floating forms existed as well. Their remains are found in both shallow- and deep-water deposits. The fusulinids were large, complex protozoans that resembled grains of wheat, ranging from microscopic size to giants up to 3 inches in length.

The earliest protists were microorganisms that built stromatolite structures. Ancestors of blue-green algae built these concentrically layered mounds, resembling cabbage heads, by cementing sediment grains together using a

Figure 18 *The contact between Manitou limestone and Precambrian gneiss and schist, Priest Canyon, Fremont County, Colorado.*

(Photo by J. C. Maher, courtesy USGS)

Figure 19 *Foraminiferans from the North Pacific Ocean.*

(Photo by R. B. Smith, courtesy USGS)

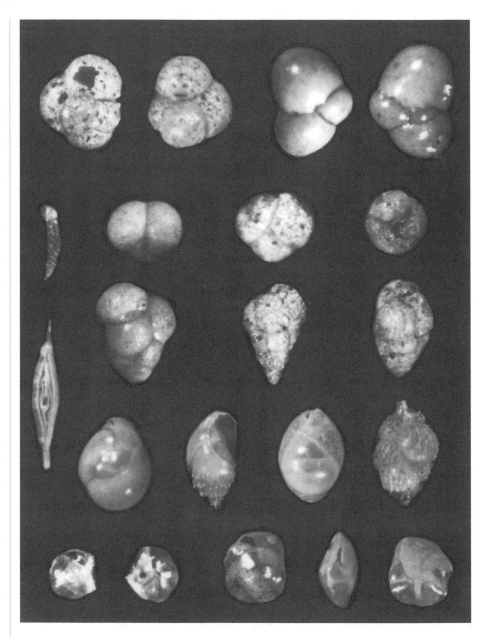

gluelike substance secreted from their bodies. The organisms ranged from the Precambrian as early as 3.5 billion years ago to the present, although many did not become well established until the Cambrian or later. Because of their lack of a hard shell, amoebas and parameciums did not fossilize well. The radiolarians, which are well represented in the fossil record, usually have siliceous shells of remarkably intricate designs, including a needlelike, rounded, or open-network structure of delicate beauty (Fig. 20).

The ability to move about under their own power is what essentially separates animals from plants, although some animals perform this function only in the larval stage and become sedentary or fixed to the seabed as adults. Mobility enabled animals to feed on plants and other animals, thus establishing new predator-prey relationships. Some organisms moved about by a thrashing tail called a flagellum, resembling a filamentous bacterium that joined the host cell for mutual benefit. Other cells had tiny hairlike appendages called cilia that propelled the organisms around by rhythmically beating the water. Many, such as the ameoba, traveled by extending fingerlike protrusions outward from the main body and flowing into them.

The earliest organisms were sulfur-metabolizing bacteria similar to those living symbiotically in the tissues of tube worms (Fig. 21). These live near sulfurous hydrothermal vents such as those on the East Pacific Rise and the Gorda Ridge off the northwest Atlantic coast of the United States. Sulfur would have been abundant on the early, hot planet, spewing from a profusion of volcanoes mostly lying on the bottom of the ocean.

Figure 20 *Late Jurassic radiolarians, Chulitna District, Alaska.*

(Photo by D. L. Jones, courtesy USGS)

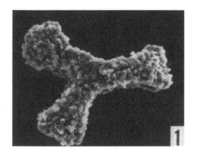 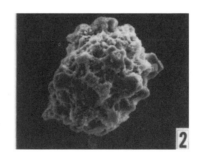 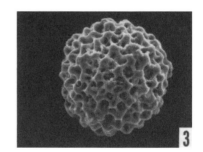

Figure 21 *Tall tube worms, giant clams, and large crabs inhabit the seafloor near hydrothermal vents.*

Sulfur combined easily with metals such as iron to form sulfates. Since the atmosphere and ocean lacked oxygen, the bacteria obtained energy by the reduction of sulfate ions. The growth of primitive bacteria was thus limited by the amount of organic molecules produced in the ocean. Although this form of energy was satisfactory at the time, bacteria were allowing a plentiful source of energy go to waste—namely sunlight.

PHOTOSYNTHESIS

The ratios of carbon isotopes in Archean rocks suggest that photosynthesis was in progress at an early age. The seas contained an abundance of iron. Oxygen generated by photosynthesis was lost by oxidation with this element, a fortunate circumstance since oxygen was also poisonous to primitive life-forms. Abundant sulfur in the early sea provided the nutrients to sustain life without the need for oxygen. Bacteria obtained energy by the reduction (the opposite of oxidation) of this important element.

Microorganisms called cyanobacteria began using sunlight as their primary energy source possibly as early as 3.5 billion years ago. The cells exploited the energy of sunlight to extract from water molecules the hydrogen they needed for self-construction, leaving oxygen as a by-product. A primitive form of photosynthesis probably began with the first appearance of blue-green algae or its predecessor, a photosynthetic bacteria called green sulfur bacteria. These organisms were best suited to an oxygen-poor environment. Oxygen was kept to a minimum by reacting with both dissolved metals in seawater and reduced gases emitted from submarine hydrothermal vents.

The first green-plant photosynthesizers called proalgae were probably intermediate between bacteria and blue-green algae. They could switch from fermentation, a primitive form of metabolism, to photosynthesis and back again, depending on their environment. Since sunlight penetrates seawater to a maximum effective depth of only a few hundred feet, the proalgae were confined to shallow water. Around 2.8 billion years ago, microorganisms called cyanobacteria began to use sunlight as their main energy source to drive the chemical reactions needed for sustained growth.

The development of photosynthesis was possibly the single most important step in the evolution of life; it gave a primitive form of blue-green algae a practically unlimited source of energy. Photosynthesis involved the absorption of sunlight by chlorophyll (in plants) to split water molecules into hydrogen and oxygen. The hydrogen combined with carbon dioxide, abundant in the early ocean and atmosphere, to form simple sugars and proteins, thereby liberating oxygen in the process (Fig. 22). The growth of photosynthetic organisms was phenomenal. The population explosion would have gotten out of hand except that oxygen, generated as a waste product of photosynthesis, was poisonous to these organisms. If not for the development of special enzymes to help organisms cope with and later use oxygen for their metabolism, life would have certainly been in jeopardy.

Photosynthesis also dramatically increased the oxygen content of the ocean and atmosphere. The oxygen level jumped significantly between 2.2 and 2 billion years ago. During that time, the ocean's high concentration of dissolved iron, which depleted free oxygen to form iron oxide, similar to the rusting of metal, was deposited onto the seafloor, creating the world's great iron ore reserves. Around this time, Earth experienced its first major period of glaciation. The cold ocean waters caused the iron to settle out of suspension.

To create and maintain an oxygen-rich atmosphere, carbon dioxide used in the photosynthetic process had to be buried in the geologic column (rock strata) in the form of carbonate rocks faster than the consumption of oxygen by the oxidation of carbon, metals, and reduced volcanic gases. About 2 billion years ago, oxygen began replacing carbon dioxide in the ocean and atmosphere. Therefore, organisms had to either develop a means of shielding their nuclei and other critical sites from oxygen or use chemical pathways that oxidized by removing hydrogen instead of adding oxygen. These innovations

Figure 22 *Energy flow in the biosphere begins with photosynthesis.*

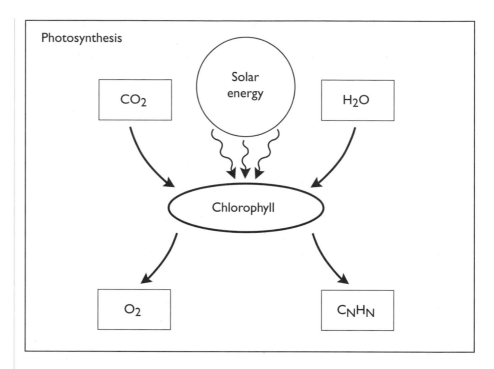

led to the evolution of the eukaryotes. Therefore, oxygen was largely responsible for the evolution of higher forms of life (Table 3).

GREENSTONE BELTS

Greenstones are ancient metamorphosed rocks unique to the Archean. Earth's crust formed in a burst of creation beginning as early as 4 billion years ago, just few million years after the formation of the planet itself. The early crust was composed of a thin layer of basalt embedded with scattered blocks of granite called "rockbergs." The granite combined into stable bodies of basement rock, upon which all other rocks were deposited. The basement rocks became the nuclei of the continents and are presently exposed in broad, low-lying, domelike structures called shields (Fig. 23).

The Precambrian shields are extensive uplifted areas surrounded by sediment-covered bedrock called continental platforms. These are broad, shallow depressions of basement complex (crystalline rock) filled with nearly flat-lying sedimentary rocks. The best-known areas are the Canadian Shield, which covers most of eastern Canada and extends down into Wisconsin and Minnesota in North America, and the Fennoscandian Shield,

TABLE 3 EVOLUTION OF LIFE AND THE ATMOSPHERE

Evolution	Origin (million years)	Atmosphere
Origin of Earth	4,600	Hydrogen, helium
Origin of life	3,800	Nitrogen, methane, carbon dioxide
Photosynthesis	2,300	Nitrogen, carbon dioxide, oxygen
Eukaryotic cells	1,400	Nitrogen, carbon dioxide, oxygen
Sexual reproduction	1,100	Nitrogen, oxygen, carbon dioxide
Metazoans	700	Nitrogen, oxygen
Land plants	400	Nitrogen, oxygen
Land animals	350	Nitrogen, oxygen
Mammals	200	Nitrogen, oxygen
Humans	2	Nitrogen, oxygen

which covers much of Scandinavia in Europe. More than a third of Australia is Precambrian shield. Sizable shields exist in Africa, South America, and Asia as well.

Many shields are fully exposed where flowing ice sheets eroded their cover of sediment during the Pleistocene ice ages. The exposure of the Canadian Shield from Manitoba to Ontario is attributed to uplifting of the crust by a mantle plume and the erosion of sediments in the uplifted area. Some of

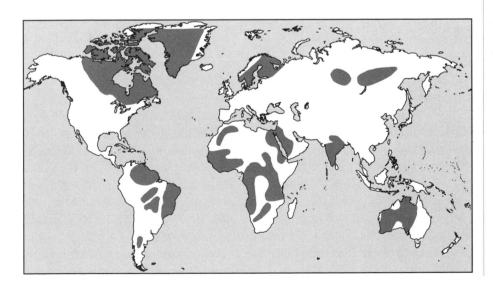

Figure 23 The Precambrian shields comprise the oldest rocks on Earth.

37

Figure 24 *Archean greenstone belts are the earliest evidence of plate tectonics.*

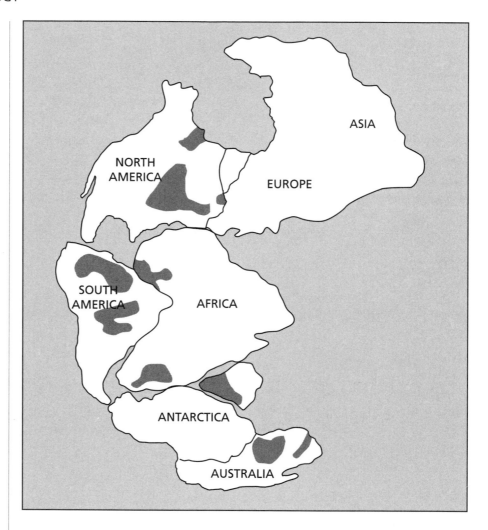

the oldest known rocks of North America are the 2.5-billion-year-old gran-ites of the Canadian Shield.

The shields are associated with greenstone belts (Fig. 24). These are a mixture of metamorphosed (recrystallized) lava flows and sediments possibly derived from island arcs (chains of volcanic islands on the edges of deep-sea trenches) caught between colliding continents. Although no large continents existed during this time, the foundations upon which they formed were pre-sent as protocontinents. These small landmasses were separated by marine basins that accumulated lava and sediments derived mainly from volcanic rocks that later metamorphosed into greenstone belts.

Greenstone belts occupy the ancient cores of the continents. They span an area of several hundred square miles, surrounded by immense expanses of gneiss, which are the metamorphic equivalents of granites and the predominant Archean rock types (Fig. 25). Their color derives from chlorite, a greenish, mica-like mineral. The existence of greenstone belts is evidence that plate tectonics (the interaction of crustal plates) might have operated as early as the Archean, some 2.7 billion years ago, with small tectonic plates clashing with each other as much as 4 billion years ago. The best-known greenstone belt is the Swaziland sequence in the Barberton Mountain Land of southeastern Africa. It is more than 3 billion years old and attains a thickness of nearly 12 miles.

Caught in the Archean greenstone belts are ophiolites, from the Greek word *ophis* meaning "serpent." They are slices of ocean floor shoved up onto the continents by drifting plates and are as much as 3.6 billion years old. Pillow lavas (Fig. 26) are tubular bodies of basalt extruded undersea. They also appear in the greenstone belts, signifying that the volcanic eruptions took place on the ocean floor. Thus, these deposits are among the best evidence that

Figure 25 *A granite pegmatite dike in Precambrian gneiss in Bear Creek Canyon, Jefferson County, Colorado.*

(Photo by J. R. Stacy, courtesy USGS)

Figure 26 *Pillow lava exposed on the south side of Wadi Jizi, northeast of Suhaylah, Sultanate of Oman.*

(Photo by E. H. Bailey, courtesy USGS)

plate tectonics operated early in the Archean. Indeed, plate tectonics appears to have been working throughout most of Earth's history in much the same manner as it does today.

The greenstone belts are of particular interest to geologists and miners because they hold most of the world's gold deposits. Archean-age ores are remarkably similar worldwide. The mineralized veins are either Archean in age or they invaded Archean rocks at a much later date. Gold of Archean age is mined on every continent except Antarctica. In Africa, the best gold deposits are in rocks as old as 3.4 billion years. Most of the South African gold mines are in greenstone belts.

The Kolar greenstone belt in India, formed when two plates crashed together about 2.5 billion years ago, holds the richest gold deposits in the world. In North America, the best gold mines are in the Great Slave region of northwest Canada, where more than 1,000 deposits are known. Most of the gold exists in greenstone belts invaded by hot magmatic solutions originating from the intrusion of granitic bodies into the greenstones. The gold occurs in veins associated with quartz.

Because greenstone belts have no equivalent in modern geology, the geologic conditions under which they formed were very different from those observed today. Active tectonic (landform-building) forces in the mantle often broke open the thin Archean crust and injected magma into the deep crustal fracture zones. Such large-scale magmatic intrusion along with numerous large meteorite impacts characterized the unusual geology of the Archean. Since greenstone belts are geologically unique to the Archean, their absence from the geologic record around 2.5 billion years ago marks the end of the eon.

ARCHEAN CRATONS

Plate tectonics has played a prominent role in shaping the planet practically from the very beginning. Continents were set adrift from the time they originated, within a few hundred million years after the formation of the planet. This is revealed by the presence of 4-billion-year-old Acasta gneiss, a metamorphosed granite, in the Slave Province of Canada's Northwest Territories. It is suggested that the formation of the crust was well underway by this time. The Slave Province contains the oldest terrestrial crust known on Earth and represents an ancient core of a continent known as a craton.

The discovery leaves little doubt that at least small patches of continental crust existed on Earth's surface during the first half-billion years of its history. The 3.8-billion-year-old metamorphosed marine sediments of the Isua Formation in a remote mountainous area in southwest Greenland (Fig. 27) provide evidence of a saltwater ocean. Earth's oldest sedimentary rocks also

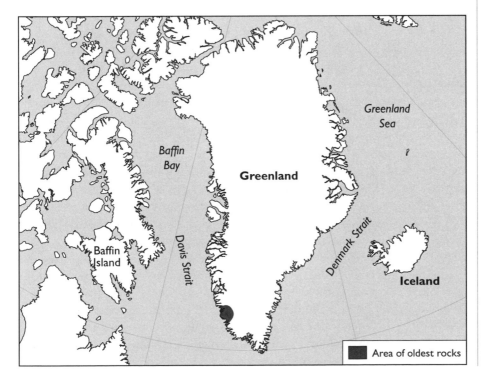

Figure 27 *The location of the Isua Formation in southwestern Greenland, which contains some of the oldest rocks on Earth.*

contain chemical fingerprints of complex cells as much as 3.9 billion years old. Around this time, a swarm of debris left over from the creation of the solar system bombarded Earth and the Moon. The bombardment might have delivered heat and organic compounds to Earth, sparking the rapid formation of primitive life. Alternatively, the pummeling could have wiped out existing life in a colossal mass extinction.

The continental crust was perhaps only about 10 percent of its present size and contained slivers of granite that drifted freely over Earth's watery face. These slices of ancient crust called Archean terranes are older than 3.3 billion years and are found notably in Canada, Greenland, southern Africa, and Western Australia. The 2.5-billion-year-old Kaapvaal, comprising a large portion of South Africa, is one of the oldest of Earth's cratons. A wedge of 3.5-billion-year-old continental crust called the Pilbara Craton, upward of thousands of square miles in length, lies in northwestern Australia and has barely been disturbed down through the ages. The area is also known for a famous formation called the Warrawoona Group that contains the world's oldest fossil cells, the earliest solid evidence of life on Earth.

Cherts are dense, extremely hard sedimentary rocks composed of microscopic grains of silica. Archean chert deposits more than 2.5 billion years old suggest that most of the crust was below the sea during this time. Most Precambrian cherts are thought to be chemical precipitates derived from silica-rich water in the deep ocean. The seas contained abundant dissolved silica, which leached out of volcanic rock pouring onto the ocean floor. Modern seawater is deficient in silica because organisms such as sponges and diatoms extract it to build their skeletons. When the organisms die, their skeletons build massive deposits of diatomaceous earth.

Few rocks date beyond 3.7 billion years, suggesting that little continental crust was being formed before this time or that the crust was recycled into the mantle. Slices of granitic crust combined into stable bodies of basement rock called cratons (Fig. 28), upon which all other rocks were deposited. They are composed of highly altered granite and metamorphosed marine sediments and lava flows. The rocks originated from intrusions of magma into the primitive ocean crust. Only three sites in the world, located in Canada, Australia, and Africa, contain rocks exposed on the surface during Earth's early history that have remained essentially unchanged throughout geologic time.

Eventually, the slices of crust began to slow their erratic wanderings and combined into larger landmasses. Constant bumps and grinds from vigorous tectonic activity built the crust inside and out. The continents continued growing rapidly up to the end of the Archean 2.5 billion years ago, when they occupied up to a quarter of Earth's surface or about 80 percent of the present continental landmass. During this time, plate tectonics began to operate extensively. Much of the world as we know it began to take shape.

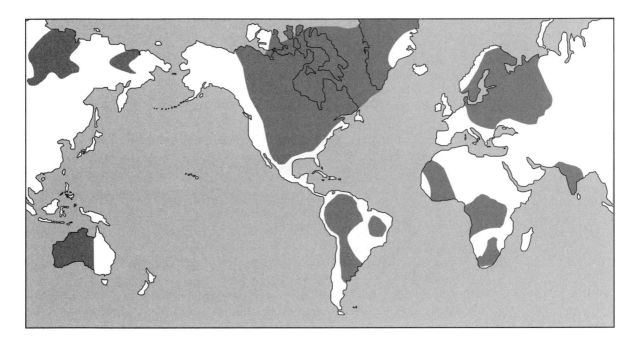

Ophiolites are the best evidence for ancient plate motions. They are sections of oceanic crust that peeled off during plate collisions and plastered onto the continents. Blueschists (Fig. 29), which are metamorphosed rocks of subducted ocean crust, also shoved up onto the continents. This resulted in a linear formation of greenish volcanic rocks along with light-colored masses of granite and gneiss, which are common igneous and metamorphic rocks that comprise the bulk of the continents. In addition, many ophiolites also contain ore-bearing rocks, which are important mineral resources throughout the world.

Figure 28 *The cratons that comprise the continents.*

The cratons were numerous. They ranged in size from about one-fifth the area of present-day North America to smaller than the state of Texas. The cratons were also highly mobile and moved about freely on the molten rocks of the upper mantle called the asthenosphere. They were independent minicontinents that periodically collided with and rebounded off each other. The collisions crumpled the leading edges of the cratons, forming small parallel mountain ranges perhaps only a few hundred feet high.

All cratons eventually coalesced into a single large landmass several thousands of miles wide called a supercontinent, which existed as early as 3 billion years ago. At the points where the cratons collided, mountain ranges pushed upward. The sutures joining the landmasses are still visible today as cores of ancient mountains called orogens, from the Greek *oros,* meaning "mountain." The original cratons formed within the first 1.5 billion years of Earth's existence and totaled only about a tenth of the present landmass.

The average rate of continental growth since then was perhaps as much as a one cubic mile a year. The constant rifting and patching of the interior along with sediments deposited along the continental margins eventually built the supercontinent outward. By the end of the Archean, it nearly equaled the total area of today's continents. Much of the landmass was covered by shallow seas until around 800 million years ago.

After discussing the simple life-forms of the Archean eon, the next chapter will focus on the more complex organisms and rock formations of the Proterozoic eon.

3

PROTEROZOIC METAZOANS

THE AGE OF COMPLEX ORGANISMS

T his chapter examines the more complex life-forms and the development of the continents during the Proterozoic eon from 2.5 billion to 545 million years ago. Major differences exist in the character of biology and geology of the Proterozoic compared with that of the Archean. The Proterozoic featured a shift to much calmer conditions as Earth progressed from a tumultuous adolescence to a stable adulthood. Marine life was distinct from species of the Archean. It represented a considerable advancement in evolution, with the development of complex organisms.

The global climate was cooler. Earth experienced its first major ice age more than 2 billion years ago along with a major extinction that eliminated many primitive species attempting to evolve during this time. The Proterozoic ended following a second period of glaciation and another mass extinction. Afterward, an explosion of species, representing nearly every major group of marine organisms, set the stage for the evolution of more modern forms of life.

THE AGE OF WORMS

Life in the Proterozoic was more advanced and complex than in the Archean. Organisms evolved very little during their first billion years on Earth because

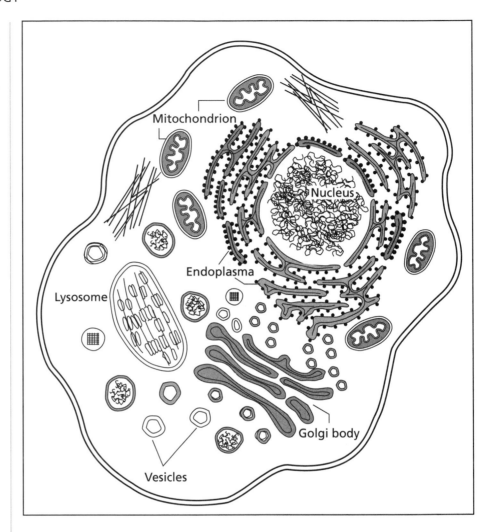

Figure 30 *The eukaryotic cell incorporates organelles along with a nucleus that contains the DNA hereditary material.*

of their primitive asexual reproduction. They used simple fission, whereby species essentially cloned themselves, which offered little evolutionary change. A primitive form of metabolism also kept organisms in a low-energy state, which retarded the rate of evolution.

The first major evolutionary advancement was the development of an organized nucleus and sexual reproduction, forming a new single-celled organism called a eukaryote (Fig. 30). It evolved from prokaryotes possibly as early as 3 billion years ago. The eukaryotic cells were typically some 10,000 times larger in volume than prokaryotes. They contained a nucleus that systematically organized genetic material, substantially increasing the number of mutations and the rate of evolution. Eukaryotes metabolized their energy sources by respiration. Therefore, their presence indicates substantial quantities of oxygen by the Proterozoic. About 2 billion years ago, when the oxygen-absorbing banded iron formations had ceased being deposited, oxygen began to replace carbon dioxide in the ocean and atmosphere.

About 1.5 billion years ago, the previously sparse geologic record of preserved cellular remains vastly improved as evolution suddenly sped up. However, nearly another billion years elapsed before multicellular animals called metazoans appeared in the fossil record. By then, the dissolved oxygen content of the sea was about 5 to 10 percent of its present value. Furthermore, the increased level of oxygen apparently sparked the evolution of many unique animals.

The triggering mechanisms for such a rapid evolutionary phase included ecologic stress, geographic isolation caused by drifting continents, and climatic changes. Organisms no longer relied entirely on surface absorption of oxygen. Gills and circulatory systems evolved when oxygen levels reached about half their present values near the end of the Proterozoic. Afterward, an explosion of species created the progenitors of all life on Earth today (Table 4).

During the latter part of the Proterozoic, around 600 million years ago, individual cells joined to form multicellular animals called metazoans. The metazoans gave rise to ever more complex organisms, which were ancestral to all

TABLE 4   CLASSIFICATION OF SPECIES

Group	Characteristics	Geologic Age
Protozoans	Single-celled animals: foraminifera and radiolarians, about 80,000 living species	Precambrian to recent
Porifera	Sponges: about 10,000 living species	Proterozoic to recent
Coelenterates	Tissues composed of three layers of cells: jellyfish, hydra, coral; about 10,000 living species	Cambrian to recent
Bryozoans	Moss animals: about 3,000 living species	Ordovician to recent
Brachiopods	Two asymmetrical shells: about 260 living species	Cambrian to recent
Molluska	Straight, curled, or two symmetrical shells: snails, clams, squids, ammonites; about 70,000 living species	Cambrian to recent
Annelids	Segmented body with well-developed internal organs: worms and leeches; about 7,000 living species	Cambrian to recent
Arthropods	Largest phylum of living species with over 1 million known: insects, spiders, shrimp, lobsters, crabs, trilobites	Cambrian to recent
Echinoderms	Bottom dwellers with radial symmetry: starfish, sea cucumbers, sand dollars, crinoids; about 5,000 living species	Cambrian to recent
Vertebrates	Spinal column and internal skeleton: fish, amphibians, reptiles, birds, mammals; about 70,000 living species	Ordovician to recent

Figure 31 *The sponges were the first giants of the sea, growing 10 feet or more across.*

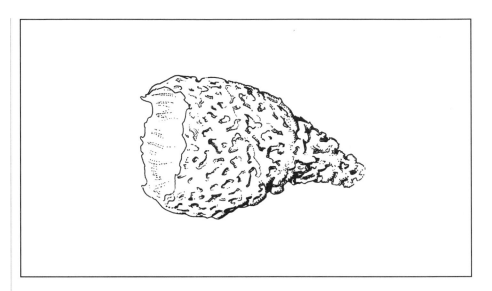

marine life today. The first metazoans were a loose organization of individual cells united for a common purpose, such as locomotion, feeding, and protection. The most primitive metazoans probably comprised numerous cells, each with its own flagellum. They grouped into a small, hollow sphere, and their flagella beat the water in unison to propel the tiny animals through the sea.

From these metazoans evolved sedentary forms turned inside out and attached to the ocean floor. Openings to the outside enabled the flagella now on the inside to produced a flow of water, providing a crude circulatory system for filtering food particles and ejecting wastes. These were the forerunners of the sponges (Fig. 31), the most primitive of metazoans. They existed in various shapes and sizes, some species possibly reaching 10 feet or more across, and grew in thickets on the ocean floor. The body consisted of three weak tissue layers whose cells could survive independently if separated from the main body. The cells either reattached themselves or grew separately into mature sponges. This ability results from the sponges' lack of regulatory genes that tell developing embryos where to put new body parts as is the case in complex animals.

Some sponges have an internal skeleton of rigid, interlocking spicules composed of calcite or silica. One group had tiny, glassy spikes for spicules, which gave the exterior a rough texture unlike their smoother relatives used in the bathtub. Glass sponges consisted of glasslike fibers of silica. These hard skeletal structures are generally the only sponge parts preserved as fossils. Sponges ranged from the Precambrian to the present, but microfossils of sponge spicules did not become abundant until the Cambrian. The great success of the sponges, which like other animals extract silica directly from seawater to construct their skeletons, explains why today's oceans are largely depleted of this mineral.

The next evolutionary step was the jellyfish, which had two layers of cells separated by a gelatinous substance, giving the saucerlike body a means of support. Unlike the cells of the sponges, those of the jellyfish were incapable of independent survival if separated from the main body. A primitive nervous system linked the cells and caused them to contract in unison, thereby providing the first simple muscles used for locomotion. Because jellyfish lacked hard body parts, they are rare as fossils and usually preserved only as carbonized films or impressions.

The development of muscles and other rudimentary organs, including sense organs and a central nervous system to process the information, followed the evolution of primitive segmented worms. The annelids are segmented worms with a body characterized by a repetition of similar parts in a long series. They include marine worms, earthworms, flatworms, and leeches. The annelids ranged from the upper Precambrian to the present. Their fossils are rare and consist mostly of tubes, tiny teeth, and jaws.

Since they were bottom dwellers, the early worms left behind a preponderance of fossilized tracks, trails, and burrows (Fig. 32) to such an extent that the Proterozoic is referred to as the "age of worms." Before about 670

Figure 32 *Fossil worm borings in Heiser Sandstone, Pensacola Mountains, Antarctica.*

(Photo by D. L. Schmidt, courtesy USGS)

Figure 33 *Stromatolite beds from a cliff above the Regal Mine, Gila County, Arizona.*

(Photo by A. F. Shride, courtesy USGS)

million years ago, however, no track-making animals existed. Marine worms burrowed into the bottom sediments or were attached to the seabed, living in tubes composed of calcite or aragonite. The tubes were almost straight or irregularly winding and attached to a solid object such as a rock, a shell, or coral. Early forms of marine flatworms grew very large, up to several feet long. The coelomic, or hollow-bodied, worms adapted to burrowing into the ocean floor sediments and might have evolved into higher forms of animal life.

Sheetlike marine worms grew nearly 3 feet long but were less than a tenth of an inch thick to provide a large surface area on which to absorb oxygen and nutrients directly from seawater. Another reason for the unusual flattened bodies of many animals was the limited food supply available during the Proterozoic. A high ratio of surface area to volume collected sunlight for algae, which lived symbiotically within their host's body. This helped to nourish their hosts while the worms supported the algae for mutual benefit.

A rounded worm from 565 million years ago was symmetrical from side to side and had what appeared to be a head. It contained three tissue layers, with an ectoderm, mesoderm, and endoderm wrapped around a gut cavity. The body apparently housed a primitive heart, a blood and vascular system, and a means of respiration. The worm left behind fossil burrows, fecal pellets, and scratches in sediments made by the very beginnings of limbs. It possibly gave rise to animals with mouths known as deuterostomes, including chordates, sea urchins, starfish, and marine worms. This ancestral creature might also have led to the protostomes, including mollusks, insects, spiders, leeches, and earthworms.

Tiny marine worms called acoels might be the closest living representatives of the first bilaterally symmetrical organisms on Earth. They are flatworms grouped along with parasites such as tapeworms and liver flukes and represent a living relic of the transition between radially symmetrical animals such as jellyfish and more complex bilateral organisms such as vertebrates and arthropods. The animals appear to be bilateral survivors from before the Cambrian, a period of unprecendented growth, providing a living window on early metazoan life. Acoels probably branched off from an ancestral animal after the radial jellyfish but before the three major bilateral groups, the vertebrates, mollusks, and arthropods, began to diverge. They therefore offer a living link between primitive and more complex animals.

The earliest fossils of plants appear to be almost entirely composed of algae that built stromatolite structures (Fig. 33). The stromatolite colonies were produced by layers of cells that excreted a gluelike substance used to bind sediments together. These were topped by photosynthetic organisms that multiplied using sunlight and supplied the lower layers with nutrients. During the late Proterozoic, around 800 million years ago, stromatolites underwent a marked decline in diversity possibly due to the appearance of algae-eating animals.

THE EDIACARAN FAUNA

Earth underwent many profound physical changes near the end of the Proterozoic, prompting a rapid radiation of species. At this time, a supercontinent called Rodinia (Fig. 34), Russian for "motherland," was located on the equator. It subsequently rifted apart, producing intense hydrothermal activity that caused fundamental environmental changes. With these profound changes in Earth, the increased marine habitat area spawned the greatest explosion of new species the world has ever known. As a result, the seas contained large

Figure 34 *The*
supercontinent Rodinia
700 million years ago.

Figure 34 *The supercontinent Rodinia 700 million years ago.*

populations of widespread and diverse organisms. Sea levels rose and flooded large portions of the land. The extended shoreline spurred the explosion of new species. Life-forms rapidly evolved, and unique and bizarre creatures roamed the ocean depths.

The dominant animals were the coelenterates. These were radially symmetrical invertebrate animals, including giant jellyfishlike floaters up to 3 feet wide and colonial feathery forms, possibly predecessors of the corals. They were attached to the ocean floor and grew more than 3 feet long. The remaining organisms were mostly marine worms, unusual naked arthropodlike animals, and a tiny curious-looking naked starfish with three rays instead of the customary five. The vase-shaped archaeocyathans (Fig. 35) resembled both sponges and corals and built the earliest limestone reefs, eventually becoming extinct in the Cambrian. Few of these creatures looked anything remotely like modern animals. The proliferation of these new organisms, representing nearly every major group of marine species, set the stage for the evolution of the progenitors of all animal life on Earth today.

The Ediacara Formation in South Australia's Ediacara Hills contains fossil impressions of highly curious creatures. The radiation of Ediacaran faunas was abrupt, caused by unprecedented environmental changes. Some Ediacaran fauna were giants in their day, reaching 3 feet in size and ranging in shape from spoked wheels and miniature anchors to corrugated and lettuce-like fronds. The Ediacaran behemoths provide the first evidence of complex life on the planet, which was previously populated by microscopic, single-celled organisms.

The complex life-forms abruptly came into existence around 600 million years ago not long after the last major episode of late Precambrian glaciation, possibly the most severe in Earth history. The emergence of the Ediacaran fauna was closely linked to profound changes in Earth's physical environment. This included the breakup of continents and the increased oxygen that made possible the evolution of large animals.

Some 30 species of these simple, beautiful creatures have been identified from fossil impressions in rock from 20 sites scattered around the world. Many impressions were left by animals apparently related to modern jellyfish, corals, and segmented worms. Others appear to represent arthropods, annelids, and possibly even echinoderms. The bodies of some forms had threefold symmetry unlike any seen in modern organisms.

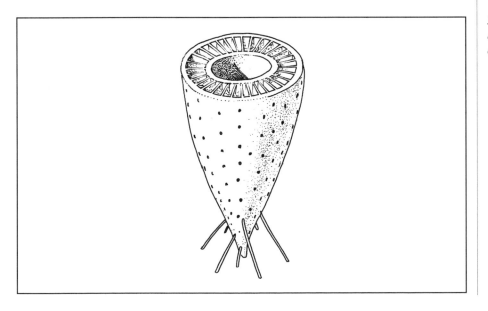

Figure 35 The archaeocyathans built the earliest limestone reefs.

The Ediacaran fossils represent a marine life very different from that of today's oceans, including feathery fronds, puckered pouches, flattened blobs, and engraved disks. Many were marked with radiating, concentric, or parallel creases, whereas others were inscribed with delicate branches. The organisms seemed to have no heads or tails; circulatory, nervous, or digestive systems; eyes; mouths; bones; muscles; or internal organs, making classification very difficult.

Ediacarans are found worldwide in rocks that just predate the Cambrian explosion, when shelly faunas first erupted onto the scene. Indeed, the disappearance of the Ediacarans made the ensuing Cambrian explosion possible. Only when the Precambrian seas were cleared of primitive life-forms could more advanced species flourish and diversify. The evolutionary heyday that marked the beginning of the Cambrian gave birth to most animal phyla that presently swim, craw, or fly around the globe. Animals appeared for the first time covered with shells, bearing jaws, claws, and other biological innovations.

The Ediacaran fauna appeared to be a preview of these animals and possessed body styles or morphologies never seen before or since in the fossil record. Most, if not all, were not related to modern forms. They were perhaps a failed evolutionary experiment in multicellular organisms completely separate from all known kingdoms of life and were wiped out in a previously unrecognized mass extinction. Yet some Ediacaran organisms apparently inhabited the planet far longer than previously thought, surviving well into the Cambrian.

The Ediacaran fossils displayed an unusual body architecture (Fig. 36) totally alien to anything seen on Earth today. Some Ediacaran faunas called

Figure 36 *The late Precambrian Ediacaran fauna from Australia.*

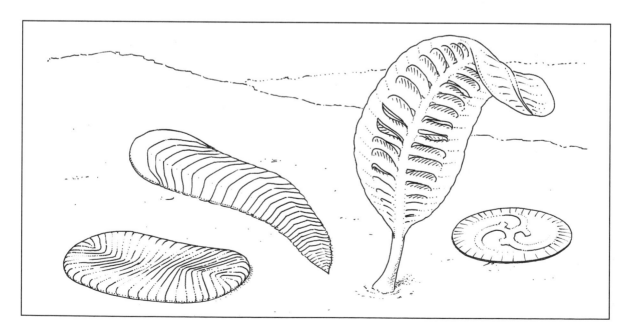

vendobionts, after the Vendian period, the final stage of the Precambrian, evolved a unique solution to the problem of growing large bodies. They simply employed networks of tubes, such as blood vessels, to transport nutrients and oxygen to individual cells. The fossils show no openings for ingesting food and eliminating wastes. They probably absorbed oxygen and nutrients directly from seawater or harbored symbiotic algae that converted sunlight into energy. They showed no obvious internal digestive or circulatory systems, which apparently did not fossilize. Nevertheless, some animals did leave what appear to be fecal pellets as a testament to an advanced digestive system.

The extremely flattened bodies of the Ediacaran faunas maximized the ratio of surface area to volume, enabling the efficient absorption of nutrients and oxygen and the collection of light for symbiotic algae. The algae lived embedded within the tissues of the host animals, which offered protection from predation in return for nutrients and the removal of waste products. These adaptations served well for the prevailing marine conditions of the late Precambrian, when shallow seas were poorly supplied with nutrients and oxygen.

The Ediacaran faunas spawned from adaptations to highly unstable conditions during the late Precambrian. Overspecialization to a narrow range of environmental conditions, however, brought about a major extinction of Ediacaran species around 540 million years ago at the very doorstep of the "Cambrian explosion," a mass diversification of species. The rapid evolution of multicellular animal life therefore ushered in the Cambrian period. Marine animals that survived the die out were markedly different from their Ediacaran ancestors and participated in the greatest diversification of new species the world has ever known. The descendants of the Ediacaran faunas split into the two great lineages of modern life, which include the protostomes such as mollusks, annelids, and arthropods, and the deuterostomes such as the echinoderms and chordates, the phylum where humans belong.

BANDED IRON FORMATIONS

Mineral deposits of the Proterozoic are bedded or stratified, as opposed to the vein deposits of the Archean. Iron, the forth most abundant element in Earth's crust, was leached from the continents and dissolved in seawater under reducing (deoxidizing) conditions. When the iron reacted with oxygen in the ocean, it precipitated (became undissolved) in vast deposits on shallow continental margins. Alternating bands of iron-rich and iron-poor sediments gave the ore a banded appearance, thus prompting the name banded iron formation or BIF. These deposits, mined extensively throughout the world, provide more than 90 percent of the minable iron reserves.

In effect, biological activity was responsible for the iron deposits since photosynthetic organisms produced the oxygen. When plants began generating oxygen (Fig. 37), it combined with iron, keeping oxygen levels in the ocean within the limits tolerated by the early prokaryotes. Throughout the Archean, the amount of oxygen probably remained less than 1 percent due to this regulating mechanism. Then, between 2.5 and 2 billion years ago, photosynthesis generated enough oxygen to react with iron on a large scale.

BIF deposits, composed of alternating layers of iron and silica, formed around 2 billion years ago at the height of the early Proterozoic ice age. For unknown reasons, major episodes of iron deposition coincided with periods of glaciation. The average ocean temperature was probably warmer than today. When warm ocean currents rich in iron and silica flowed toward the glaciated polar regions, the suddenly cooled waters could no longer hold minerals in solution. As they precipitated out of seawater, they formed alternating layers on the ocean floor due to the difference in settling rates between silica and iron, the heavier of the two minerals. After most of the dissolved iron was

Figure 37 The evolution of life and oxygen in the atmosphere.

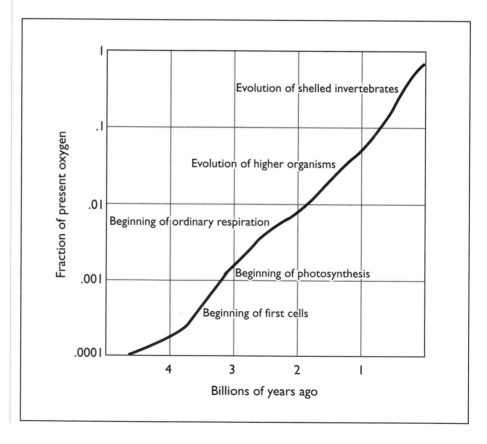

Figure 38 Metal-rich massive sulfide vein deposit in ophiolite.

(Photo courtesy USGS)

locked up in the sediments, the level of oxygen began to rise steadily, spawning the evolution of more advanced species.

Biochemical activity in the ocean was also responsible for stratified sulfide deposits (Fig. 38). Sulfur-metabolizing bacteria living near undersea hydrothermal vents oxidized hydrogen sulfide into elemental sulfur and various sulfates. Copper, lead, and zinc, which were much more abundant in the Proterozoic than in the Archean, also reflect a submarine volcanic origin.

PRECAMBRIAN GLACIATION

The Proterozoic was a period of transition, when oxygen generated by photosynthesis replaced carbon dioxide. Early in the Archean, the Sun's output was only about 70 percent of its present value. Large amounts of atmospheric carbon dioxide 1,000 times greater than current levels kept Earth from freezing solid. When the first microscopic plants evolved, they began replacing carbon dioxide in the ocean and atmosphere with oxygen so that today the relative percentages of these two gases have completely reversed. The loss of carbon dioxide, an important greenhouse gas, caused the climate to cool even while the Sun was getting progressively hotter.

The drop in global temperatures soon after the beginning of the Proterozoic initiated the first known glacial epoch about 2.4 billion years ago (Table 5)

TABLE 5   CHRONOLOGY OF THE MAJOR ICE AGES

Time (years)	Event
10,000–present	Present interglacial
15,000–10,000	Melting of ice sheets
20,000–18,000	Last glacial maximum
100,000	Most recent glacial episode
1 million	First major interglacial
3 million	First glacial episode in Northern Hemisphere
4 million	Ice covers Greenland and the Arctic Ocean
15 million	Second major glacial episode in Antarctica
30 million	First major glacial episode in Antarctica
65 million	Climate deteriorates, poles become much colder
250–65 million	Interval of warm and relatively uniform climate
250 million	The great Permian ice age
700 million	The great Precambrian ice age
2.4 billion	First major ice age

---

when massive sheets of ice nearly engulfed the entire landmass. The positions of the continents also had a tremendous influence on the initiation of ice ages. Lands wandering into the colder latitudes enabled the buildup of glacial ice. Global tectonics, featuring extensive volcanic activity and seafloor spreading, might have triggered the ice ages by drawing down the level of oxygen in the ocean and atmosphere. This would have preserved more organic carbon in the sediments so that living organisms could not return it to the atmosphere.

Plate tectonics also began to operate more vigorously. Plate subduction thrust carbonaceous sediments deep inside Earth along with the underlying oceanic crust. The growing continents stored large quantities of carbon in thick deposits of carbonaceous rocks such as limestone. The elimination of carbon dioxide in this manner caused Earth to cool dramatically. Besides high rates of organic carbon burial, iron deposition and intense hydrothermal activity associated with plate tectonics furthered global cooling. Although this was the first ice age the world had ever known, it was not the worst.

The burial of carbon in Earth's crust might have been the key to the onset of another glacial period just before the appearance of recognizable animal life near the end of the Proterozoic about 680 million years ago, called the Varanger ice age, named for the Varanger Fjord in Norway. Massive glaciers overran nearly half the continents for millions of years. Four periods of glaciation occurred between 850 and 580 million years ago. Even the tropics froze over, as indicated by the occurrence of glacial debris and unusual deposits of iron-rich rocks on virtually every continent, indicating they formed in cold waters. The ice age nearly snuffed out all life, which endured extremely hard living conditions for millions of years.

A supercontinent located on the equator rifted apart, forming four or five major continents. The largest apparently wandered into the southern polar region and acquired a thick blanket of ice. This was perhaps the greatest and most prolonged period of glaciation when ice encased nearly half the planet. The ice cover was so extensive that the period has been dubbed the *snowball Earth*. If not for massive volcanic activity, which restored the carbon dioxide content of the atmosphere, the planet could still be buried under ice.

The climate was so cold that ice sheets and permafrost (permanently frozen ground) extended into equatorial latitudes. During this time, no plants grew on the land, and only simple plants and animals lived in the sea. The ice age dealt a deathblow to life in the ocean, and many simple organisms vanished during the world's first mass extinction. At this point in Earth's development, animal life was still scarce. The extinction decimated the ocean's population of acritarchs, a community of planktonic algae that were among the first organisms to develop elaborate cells with nuclei. These extreme glaciations took place just before a rapid diversification of multicellular life, culminating in an explosion of species between 575 and 525 million years ago.

The glacial periods are marked by deposits of moraines and tillites. Thick sequences of Precambrian tillites exist on every continent (Fig. 39). Tillites are a mixture of boulders and clay deposited by glacial ice and cemented into solid rock. In the Lake Superior region of North America, tillites are 600 feet thick in places and range from east to west for 1,000 miles. In northern Utah, tillites mount up to 12,000 feet thick. The various layers of glacially deposited sediment suggest a series of ice ages closely following each other. Similar tillites are among Precambrian rocks in Norway, Greenland, China, India, southwest Africa, and Australia.

Australia was practically buried by glaciers during the late Precambrian ice age (Fig. 40). The 680-million-year-old glacial varves in lake bed deposits north of Adelaide, South Australia, might hold evidence of a solar cycle operating in the Proterozoic. The varves consist of alternating layers of silt laid down annually. Each summer when the glacial ice melted, sediment-laden meltwater discharged into a lake, and sediments settled out in a stratified deposit. During times of intense solar activity, average global temperatures increased. As a result, the glaciers melted more rapidly, depositing thicker varves. By counting the layers of thick and thin varves, a stratigraphic sequence is established that mimics both the 11-year sunspot cycle and the 22-year solar cycle or possibly the early lunar cycle, which today is about 19 years.

Rocks containing carbonate minerals lie just above the glacial deposits. This indicates a hothouse effect produced by a profusion of volcanic activity that pumped carbon dioxide into the atmosphere, heating Earth and melting the great glaciers. Not long after the ice disappeared near the end of the Proterozoic, a great diversity of animal life culminated with the evolution of entirely new species, the likes of which have never existed before or since. This forever changed the composition of Earth's biology. Life-forms took off in all directions, producing many unique and bizarre animals. The rapid evolution produced three times as many phyla, groups of organisms sharing the same general body plan, as are living today.

Many unusual creatures occupied the ocean. The highest percentage of experimental organisms, animals that evolved special characteristics, came into being than during any other interval of Earth history. As many as 100 phyla existed, whereas only about a third that number are living today. This biological exuberance set the stage for more advanced species. Also, for the first time, fossilized remains of animals became abundant due to the evolution of lime-secreting organisms that constructed hard shells.

Figure 39 The location of Precambrian glacial deposits.

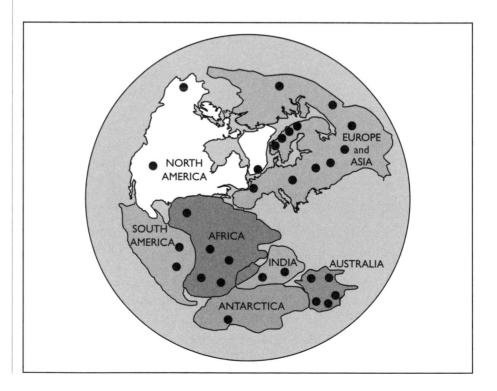

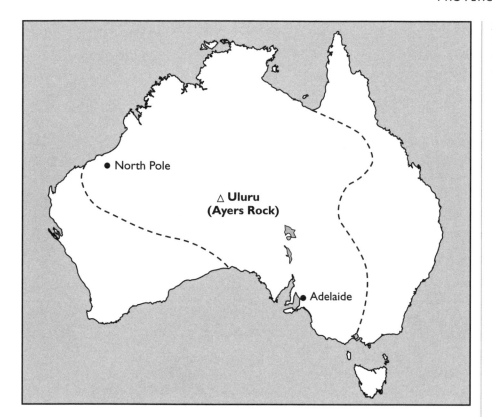

Figure 40 *The dashed lines indicate the extent of the late Precambrian ice age in Australia.*

North Pole

△ **Uluru (Ayers Rock)**

● Adelaide

THE CONTINENTAL CRUST

By the beginning of the Proterozoic, upward of 80 percent or more of the present continental crust was already in existence (Fig. 41). Most of the material presently locked in sedimentary rocks was at or near the surface. Therefore, ample sources of Archean rocks were available for erosion and redeposition. Sediments derived directly from primary sources are called graywackes, often described as dirty sandstone. They commonly occur in folded sedimentary rocks such as those in the Alps and Alaska. Most Proterozoic graywackes composed of sandstones and siltstones originated from Archean greenstones. Another common rock type was a fine-grained quartzite. It was a metamorphic rock derived from the erosion of siliceous grainy rocks such as granite and course-grained sandstone called arkose.

Conglomerates, consolidated equivalents of gravels, were also abundant in the Proterozoic. Nearly 20,000 feet of Proterozoic sediments form the Uinta Range of Utah (Fig. 42), and the Montana Proterozoic belt system contains sediments more than 11 miles thick. The Proterozoic is also

Figure 41 *Sawatch sandstone resting on Precambrian granite, Ute Pass, El Paso County, Colorado.*

(Photo by N. H. Darton, courtesy USGS)

known for its widespread terrestrial redbeds, named so because sediment grains were cemented with red iron oxide. Their appearance around 1 billion years ago indicates that the atmosphere had substantial levels of oxygen at this time.

The weathering of primary rocks produced solutions of calcium carbonate, magnesium carbonate, calcium sulfate, and sodium chloride, which subsequently precipitated into limestone, dolomite, gypsum, and halite. Higher carbon dioxide concentrations in the Precambrian probably account for the prevalence of dolomite over limestone. These minerals appear to be mainly chemical precipitates and not of biological origin.

In the Mackenzie Mountains of northwest Canada, dolomite deposits range up to 6,500 feet thick. In the Alps, massive chunks of dolomite soar skyward. Carbonate rocks such as limestone and chalk, produced chiefly by organic processes involving shells and skeletons of simple organisms, became more common in the late Proterozoic beginning about 700 million years ago. In contrast, they were relatively rare in the Archean due to the scarcity of lime-secreting organisms.

The continents of the Proterozoic comprised Archean cratons. The North American continent assembled from seven cratons around 2 billion years ago, making it the oldest continent. The cratons welded into what is now central Canada and north-central United States. Meanwhile, continental collisions continued to add a large area of new crust to the growing proto–North American continent (Fig. 43). At Cape Smith on Hudson Bay is a 2-billion-year-old slice of oceanic crust that was accreted onto the land, indicating that

Figure 42 *Ladore Canyon and the Green River in the Uinta Mountains, Summit County, Utah.*

(Photo by W. R. Hansen, courtesy USGS)

Figure 43 The proto–North American continent continued to grow by adding continental fragments to its landmass.

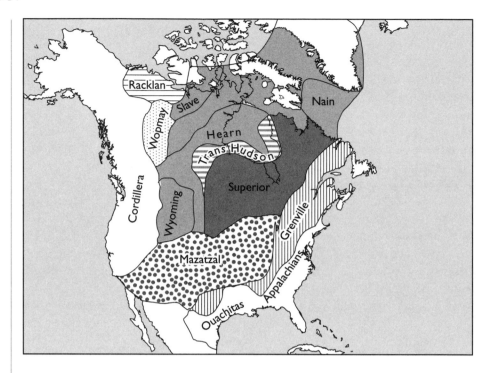

continents collided and closed an ancient ocean. Arcs of volcanic rock also weave through central and eastern Canada down into the Dakotas.

A region between Canada's Great Bear Lake and the Beaufort Sea holds the roots of an ancient mountain range that formed by the collision of North America and an unknown landmass between 1.2 and 0.9 billion years ago. A continental collision possibly with South America created the Grenville orogeny between 1.3 and 0.9 billion years ago, which formed a 3,000-mile-long mountain belt in eastern North America from Texas to Labrador. Meanwhile, continental collisions continued to add a large area of new crust to the growing proto–North American continent.

Most of the continental crust underlying the United States from Arizona to the Great Lakes to Alabama formed in one great surge of crustal generation unequaled in North America since. The rapid buildup of new crust resulted from possibly the most energetic period of tectonic activity than any subsequent time in Earth history. The assembled North American continent was stable enough to resist a billion years of jostling and rifting. It continued to grow by adding bits and pieces of continents and island arcs to its edges.

Large masses of volcanic rock near the eastern margin of North America imply that the continent was the core of a larger supercontinent called Rodinia formed during the late Proterozoic. The interior of this landmass heated and erupted with volcanism. The warm, weakened crust consequently broke apart into several major continents between 630 and 560 million years ago, although their shapes differed from the continents of today. The breakup of the supercontinent created thousands of miles of new continental margin, which might have played a major role in the explosion of new life at the beginning of the Phanerozoic.

After covering life in the Proterozoic, the next chapter takes a look at the invertebrate life-forms and rocks of the Cambrian period.

4

CAMBRIAN INVERTEBRATES
THE AGE OF SHELLY FAUNAS

This chapter examines the invertebrate life-forms and landforms of the lower Paleozoic era. The Cambrian period from 545 to 500 million years ago was named for a mountain range in central Wales, Great Britain that contained sediments with the earliest known fossils. Nineteenth-century geologists were often puzzled why ancient rocks were practically devoid of fossils until the Cambrian, when life suddenly sprang up in great abundance the world over. The base of the Cambrian (Fig. 44) was therefore thought to be the beginning of life, and all time before then was simply called Precambrian.

The period was generally quiet in terms of geologic processes, with little mountain building, volcanic activity, glaciation, and extremes in climate. The breakup of the late Precambrian supercontinent Rodinia and the flooding of continents with inland seas created abundant, warm, shallow-water habitats, prompting an explosion of new species. For more than 80 percent of its history, life remained almost exclusively unicellular. Then less than 600 million years ago, multicellular organisms abruptly came onto the scene. Never before or since had so many novel and unusual organisms existed; surprisingly, none have any counterpart in today's living world.

THE CAMBRIAN EXPLOSION

The Cambrian was an evolutionary heyday for species, featuring a 10-million-year rapid evolution of the first complex animals with exoskeletons. The Cambrian witnessed the first appearance of nearly all animal phyla in the fossil record. After 3 billion years of unicellular life, a short term of intense creativity was followed by more than 500 million years of variations on anatomic themes established at the beginning of the Cambrian. So many new and varied life-forms came into existence that the age is depicted as the Cambrian explosion.

The Cambrian explosion was the most remarkable and puzzling event in the history of life. The early Cambrian witnessed the disappearance of soft-bodied Ediacaran faunas and the rapid proliferation of shelly faunas (Fig. 45). The biological proliferation peaked about 530 million years ago, filling the ocean with a rich assortment of life. Seemingly out of nowhere and in bewildering abundance, animals appeared in an astonishingly short time span cloaked in a baffling variety of exoskeletons.

Figure 44 *A syncline in weathered shale of the lower Cambrian Rome Formation, Johnson County, Tennessee.*

(Photo by W. B. Hamilton, courtesy USGS)

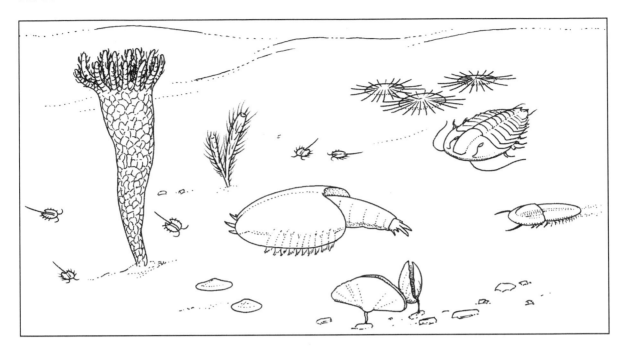

Figure 45 *Early Cambrian marine fauna.*

The introduction of hard skeletal parts has been called the greatest discontinuity in Earth history. It signaled a major evolutionary change by accelerating the developmental pace of new organisms. Nearly all major groups of modern animals appeared in the fossil record. For the first time, animals sported shells, skeletons, legs, and sensing antennae. Furthermore, a stable environment enabled marine life to flourish and disperse to all parts of the world.

The period follows on the heels of the great Precambrian ice age, the worst the world has ever experienced, when ice sheets covered half the world. It was also a time when oxygen concentrations rose to significant levels. After the ice retreated and the seas began to warm, life took off in all directions. The Cambrian saw the highest percentage of experimental organisms than any other interval of geologic history, with perhaps three times more phyla in existence than today.

At the beginning of the Cambrian, an ocean turnover might have brought unusual amounts of nutrient-rich bottom water to the surface. Due to an increase in seawater calcium levels, early soft-bodied creatures developed skeletons as receptacles for the disposal of excess amounts of this toxic mineral from their tissues. As concentrations of calcium in the ocean further increased, animal skeletons became more diverse and elaborate.

Levels of atmospheric oxygen appeared to rise in concert with the skeletal revolution. The higher oxygen levels increased metabolic energy, enabling the growth of larger animals. These, in turn, required stronger structural supports. Skeletons also evolved as a response to an incoming wave of fierce predators. Paradoxically, most of these predators were soft-bodied and therefore not well preserved as fossils.

Soft-bodied organisms living before the arrival of shelled animals at the beginning of the Cambrian had a great difficulty entering the fossil record. Animals with soft body parts decayed rapidly upon death. Therefore, only traces of their existence remain, such as imprints, tracks, and borings. These include fossil impressions of soft-bodied animals found in the Ediacara Hills of southern Australia that date from about 670 million years ago.

The Ediacaran fauna spawned from adaptations to highly unstable conditions during the late Precambrian. Overspecialization to a narrow range of environmental conditions caused a major extinction of Ediacaran species around 540 million years ago at the very beginning of the Cambrian explosion. Marine animals that survived the die out differed markedly from their Ediacaran ancestors and participated in the greatest diversification of new species the world has ever known.

At the dawn of the Cambrian, 125 million years following the appearance of the Ediacaran fauna, most of which were evolutionary dead ends, the seascape abruptly changed with the sudden arrival of animals with hard skeletal parts. Most phyla of living organisms appeared almost simultaneously, many of which had their origins in the latter part of the Precambrian. Body styles that evolved in the Cambrian largely served as blueprints for modern species, with few new radical body plans appearing since then.

When skeletons evolved, the number of organisms preserved in the fossil record jumped dramatically. All known animal phyla that readily fossilized appeared during the Cambrian, after which the number of new classes decreased sharply. Fossils became abundant at the beginning of the period because of the development of hard body parts that fossilize by replacement with calcium carbonate or silica, rapid burrowing to prevent attack by scavengers and decay by oxidation, long periods of deposition with little erosion, and large populations of species.

THE AGE OF TRILOBITES

The Cambrian is best known for a famous group of invertebrates called trilobites (Fig. 46). Early in the Cambrian, a wave of extinctions decimated a huge variety of newly evolved species. The mass extinction eliminated more than 80 percent of all marine animal genera and is numbered among the worst in Earth history. It coincided with a drop in sea level that followed continental collisions. The die-offs paved the way for the ascendancy of the trilobites,

Figure 46 *Trilobites, extinct ancestors of today's horseshoe crabs, first appeared in abundance in the Cambrian and became extinct at the close of the Paleozoic.*

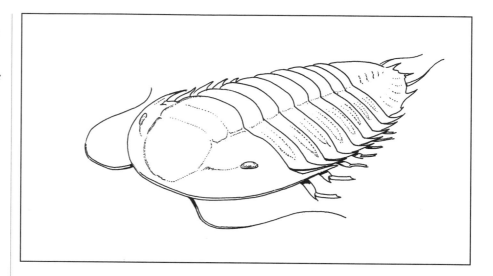

which were among the first organisms to grow hard shells and the dominant species for the next 100 million years.

Trilobites were primitive aquatic crustaceans and a favorite among fossil collectors, second only to dinosaurs as fossils of public interest. They were as common and diverse as their crustacean cousins are today. Because most of the present world's land areas were submerged during various parts of the Paleozoic, trilobites are found in marine sedimentary rocks on all continents. They lived mostly on the seafloor and were among the earliest known organisms to grow hard shells. The remarkable structure and beauty of the trilobite compound eye provides important information about visual capabilities early in the history of invertebrates.

The trilobites were small, oval-shaped arthropods, ancestors to the horseshoe crab, their only remaining direct descendent. Their body had three lobes, hence the name. It consisted of a central axial lobe containing the creature's essential organs and two side, or pleural, lobes. The giant paradoxides were truly a paradox among trilobites, extending nearly 2 feet in length, while most trilobites were generally less than 4 inches long. Since trilobites were so widespread and lived throughout the Paleozoic, their fossils are important markers for dating rocks of this era. They appeared at the very base of the Cambrian and became the dominant invertebrates of the early Paleozoic. They diversified into about 10,000 species before declining and becoming extinct after some 300 million years of highly successful existence.

The trilobites occupied the shallow waters near the shores of ancient seas that flooded inland areas, providing extensive continental margins from the coastline to the edge of the abyss. Seawater advanced to cover most of the continents

540 million years ago, as evidenced by the presence of Cambrian seashores in such places as the interior of North America, which is remarkably similar on all continents. In addition, a stable environment enabled marine life to flourish and proliferate, becoming widely dispersed. The flora included cyanobacteria, red and green algae, and acritarchs, a form of plankton that supported early Paleozoic food chains. Some simple multicellular algae that built algal mats and stromatolites similar to those of today evolved more than a billion years ago.

Curiously, many trilobite fossils have bite scars predominantly on the right side. Predators might have attacked from the right probably because when the trilobite curled up to protect itself it exposed this side of its body. (Trilobite fossils are often found with their bodies completely curled up.) However, if the trilobite had a vital organ on its left side and an attack occurred there, it stood a good chance of being eaten, thereby leaving no fossil. Therefore, if attacked on the right side, the trilobite stood a better chance of entering the fossil record, albeit with a chunk bitten out of its side.

Trilobites shed their exoskeletons as they grew. In this manner, an individual could have left several fossils, which explains why whole fossils are rare (Fig. 47). During molting, a suture opened across the head, and the trilobite simply fell out of its exoskeleton. Sometimes, though, a clean suture failed to open, and the animal had to wiggle its way out. Either way, it was still vulnerable to predators until its new skeleton hardened.

The population of trilobites peaked during the late Cambrian around 520 million years ago, when they accounted for about two-thirds of all marine species. By about 475 million years ago, the number of trilobite species abruptly declined to one-third following the rise of mollusks, corals, and other stationary filter feeders. Later, the trilobites left the nearshore areas for the offshore probably in response to significant environmental changes in temperature and seawater chemistry. The demise of the trilobites also appears to follow the arrival of the jawed fish.

CAMBRIAN PALEONTOLOGY

The sponges were primitive multicellular animals that thrived on the seafloor just before the start of the Cambrian. The early sponges lived at a pivotal point in Earth history, when ancestors of most modern animal groups suddenly appeared in the seas. The ancient sponges lived much as their modern counterparts do. They filtered nutrients from seawater, thus providing the earliest example of filter feeding, which is an extremely common mode of life in the modern seas.

The existence of sponges prior to the Cambrian suggests that more complex animals must have evolved by the time the Ediacaran fauna appeared around the world. The sponge skeletons consisted of tiny glassy spikes very sim-

Figure 47 Trilobite fossils of the Cambrian age Carrara Formation in the southern Great Basin of California and Nevada.

(Photo by A. R. Palmer, courtesy USGS)

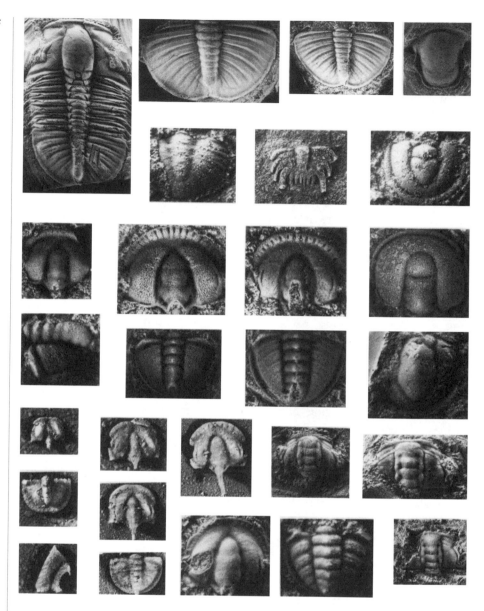

ilar to the spicules of some sponges today, indicating little change had taken place over the intervening period unlike in most other groups. The glass-barbed sponges, however, are quite different from the types used in the bathtub.

The coelenterates, from Greek meaning "gut," were well represented in the Cambrian seas. They are the most primitive of animals and include jellyfish, sea anemones, sea pens, and corals. Most coelenterates are radially symmetrical, with body parts radiating outward from a central point. They have a saclike body and a mouth surrounded by tentacles. Paleontologists have yet to find animal fossils in rocks more than 600 million years old. However, they have discovered the oldest known animal embryos of jellyfish preserved within tiny fossilized eggs from the early Cambrian. The discovery indicates that evidence for billion-year-old complex animals might be found by searching for embryos in ancient rocks.

Primitive, radially symmetrical animals have just two types of cells, the ectoderm and endoderm. In contrast, bilaterally symmetrical animals also have a mesoderm and a distinct gut. During early cell division called cleavage in bilateral animals, the fertilized egg forms two and then four cells, each of which produces several small cells. Many types have two stages of development, consisting of a stationary polyp attached to the seabed by its base with tentacles and mouth directed upward, and a mobile umbrella-shaped medusa or jellyfishlike phase with tentacles and mouth directed downward.

The corals possess a large variety of skeletal forms (Fig. 48), and successive generations built thick limestone reefs. Corals began constructing reefs in the lower Paleozoic, forming chains of islands and barrier reefs along the shorelines of the continents. The archaeocyathans resembled both corals and sponges. However, they have no close relationship to any living group and thus belong in their own unique phylum. These cone-shaped creatures formed the earliest reefs, eventually becoming extinct in the Cambrian. Many corals declined and were replaced by sponges and algae in the late Paleozoic due to the recession of the seas in which they once thrived.

The coral polyp is a soft-bodied creature that is essentially a contractible sac crowned by a ring of tentacles. The tentacles surround a mouthlike opening and are tipped with poisonous stingers. The polyps live in individual skeletal cups or tubes called thecae composed of calcium carbonate. They extend their tentacles to feed at night and withdraw into their thecae during the day or at low tide to keep from drying in the Sun. The corals coexist symbiotically (in conjunction) with zooxanthellae algae, which live within the polyp's body. The algae consume the coral's waste products and produce organic materials that nourish the polyp. Some coral species receive 60 percent of their food from algae. Because the algae require sunlight for photosynthesis, corals must live in warm, shallow water generally less than 100 feet deep and at temperatures between 25 and 30 degrees Celsius.

The echinoderms, from Greek meaning "spiny skin," were perhaps the strangest animals ever preserved in the fossil record of the early Paleozoic. Their fivefold radial symmetry with arms radiating outward from the center of the body makes them unique among the more complex animals. They are the only species possessing a water vascular system composed of internal canals that operated a series of tube feet, or podia, used for locomotion, feeding, and respiration. The great success of the echinoderms is illustrated by the fact that

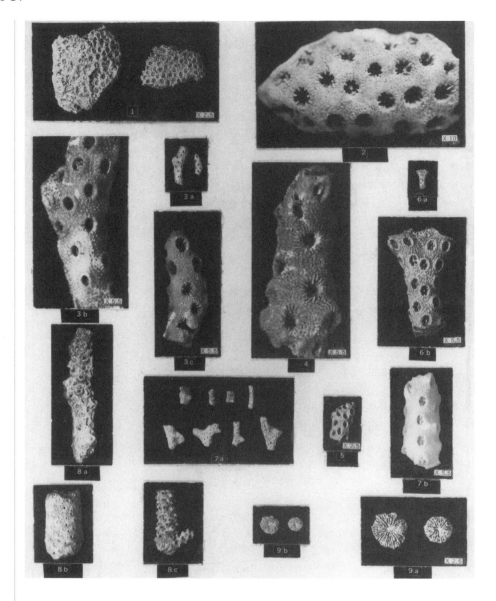

more classes of this animal exist than of any phylum both living and extinct. The major groups of echinoderms include starfish, brittle stars, sea urchins, sea cucumbers, and crinoids (Fig. 49), known as sea lilies because of their plant-like appearance.

Surprisingly, according to genetic studies, early echinoderms and chordates were more closely related to each other than to other major phyla. Apparently, echinoderms and chordates branched away from the arthropods, annelids, and mollusks as early as 1.2 billion years ago. This was long before the Cambrian explosion, when almost every major group of animals abruptly appeared in the fossil record. About 1 billion years ago, echinoderms and chordates went their separate ways. This suggests that a slow-paced course of evolution was well under way half a billion years prior to the Cambrian explosion.

The brachiopods (Fig. 50), also called lampshells because of their resemblance to ancient oil lamps, were once the most abundant and diverse organisms. More than 30,000 species are cataloged in the fossil record. The prolific brachiopods ranged from the Cambrian to the present but were most abundant in the Paleozoic and to a lesser extent in the Mesozoic. The appearance of brachiopod fossils in the strata indicates that seas of moderate to shallow depth once covered the area. The brachiopods had two saucerlike shells called valves that fitted face to face and opened and closed using simple muscles.

Brachiopods filtered food particles through opened shells that closed to protect the animals against predators. More advanced species, including brachiopods called articulates, had ribbed shells with interlocking teeth that opened and closed along a hinge. Many brachiopods are excellent index fossils for correlating rock formations throughout the world. They are important as guide fossils and are used to date many Paleozoic rocks.

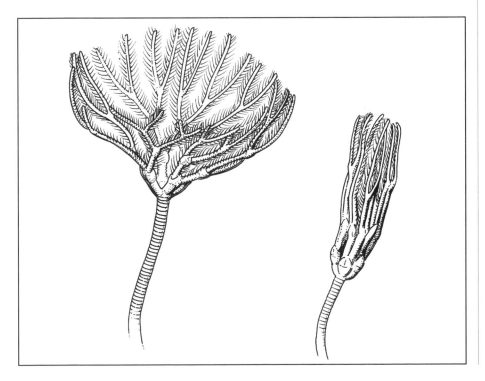

Figure 49 Crinoids were a dominate species in the middle and late Paleozoic and are still in existence today.

75

Figure 50 *Fossil casts of brachiopods.*

(Photo by E. B. Hardin, courtesy USGS)

The mollusks are a highly diverse group and left the most impressive fossil record of all marine animals (Fig. 51). They make up the second largest of the 21 basic animal groups, behind the arthropods. The phylum is so diverse that paleontologists have difficulty finding common features among its members. The three major groups are the snails, clams, and cephalopods. Snails and slugs comprise the largest group and ranged from the Cambrian onward. An ancient mollusk called kimberella (Fig. 52) was an odd Precambrian creature shaped like a gherkin cucumber and lived just before the Cambrian explosion.

The cephalopods, which include the squid, cuttlefish, octopus, and nautilus, traveled by jet propulsion. They sucked water into a cylindrical cavity through openings on each side of the head and expelled it under pressure through a funnel-like appendage. Their straight, streamlined shells, up to 30 feet and more in length, made the nautiloids among the swiftest animals of the ancient seas. The ammonoids were the most spectacular of marine predators, with a large variety of coiled shell forms.

The arthropods are the largest phylum of living organisms, comprising roughly 1 million species, or about 80 percent of all known animals. The giant 3-foot-long arthropods found in the middle Cambrian Burgess Shale Formation of western Canada represent one of the largest of all Cambrian invertebrates. Among the first and best known of the ancient arthropods were the trilobites. The arthropod body is segmented, suggesting a relationship to the annelid worms. Paired, jointed limbs generally present on most segments are modified for sensing, feeding, walking, and reproduction.

The crustaceans appeared at the very beginning of the Cambrian and soon became the dominant invertebrates. They occupied shallow waters near the shores of ancient seas, which flooded inland to provide extensive continental margins. The crustaceans are primarily aquatic arthropods and include shrimps, lobsters, barnacles, and crabs (Fig. 53). The ostracods, or mussel shrimps, are small crustaceans found in both marine and freshwater environments. Their fossils are useful for correlating rocks from the early Paleozoic onward, which make them particularly important to geologists.

The conodonts are fossilized, tiny, jawlike appendages that commonly occur in marine rocks ranging from the Cambrian to the Triassic and are important for dating Paleozoic rocks. They are among the most baffling of all fossils and have puzzled paleontologists since the 1800s. Paleontologists began finding these isolated toothlike objects in rocks from the late Cambrian

Figure 51 *Molds and shells of mollusks on highly fossiliferous sandstone of the Glenns Ferry Formation on Deadman Creek, Elmore County, Idaho.*

(Photo by H. E. Malde, courtesy USGS)

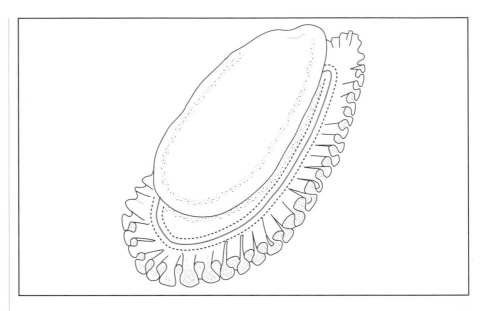

Figure 52 *Kimberella was an odd-looking Precambrian mollusk.*

through the Triassic periods. The conodonts are thought to be bony appendages of an unusual, soft-bodied animal resembling a hagfish. However, the shape of the missing creature remained unclear until 1983, when paleontologists discovered toothlike pieces at the forward end of an eel-shaped fossil from Scotland. Furthermore, the presence of distinctive eye muscles in conodonts not known in invertebrates pushed the vertebrate fossil record as far back as the Cambrian. Conodonts show their greatest diversity during the Devonian and are important for long-range rock correlations of that period.

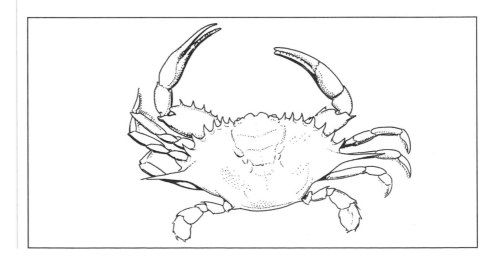

Figure 53 *Crustaceans such as this crab are primarily aquatic species.*

Graptolites were colonies of cupped organisms that resembled stems and leaves of plants but were actually animals, with individual organisms housed in tiny cups. They clung to the seafloor like small shrubs, floated freely near the surface, appearing much like tiny saw blades, or attached to seaweed. Large numbers of graptolites buried in the bottom mud produced organic-rich black shales that indicate poor oxygen conditions. They are important markers for correlating rock units of the lower Paleozoic.

THE BURGESS SHALE FAUNA

The Burgess Shale Formation in British Columbia, Canada, contains the remains of bizarre, soft-bodied animals that first appeared in the lower Cambrian about 540 million years ago soon after the emergence of complex organisms. The assemblage featured more diversity of basic anatomic designs than all the world's oceans today. It included some two dozen types of plants and animals that have no modern counterparts. Many of these peculiar animals were possibly carried over from the upper Precambrian but never made it beyond the middle Paleozoic. They were so strange that they defied efforts to classify them into existing taxonomic groups.

The animals were surprisingly complex with specialized adaptations for living in a variety of environments. Some species appear to be surviving Ediacaran fauna, most of which became extinct near the end of the Precambrian. Indeed, the Cambrian explosion might have been triggered in part by the availability of habitats vacated by departing Ediacaran species. The Burgess Shale fauna comprised more than 20 distinct body plans. A leechlike animal called *Pikaia* (Fig. 54) was the first known member of the chordate phylum and one of the rarest fossils of the Burgess Shale Formation, which contains the best-preserved Cambrian fauna. It had a stiff dorsal rod called a notochord along the back made of cartilage that supported organs and muscles, the predecessor of the spine in vertebrates.

The Burgess Shale fauna originated in shallow water on a gigantic coral reef that dwarfed Australia's Great Barrier Reef, the single most massive structure built by present-day living beings. The ancient reef surrounded Laurentia, the ancestral North American continent, and was covered with mud that readily trapped and fossilized organisms. Most occurrences originated from the western Cordillera of North America, an ancient mountain range that faced an open ocean in the middle Cambrian. Similar faunas existed on other cratons, including the North and South China blocks, Australia, and the East European platform. Their widespread distribution around other continents suggests many members had a swimming mode of life.

Figure 54 Pikaia *was the first known member of the chordate phylum.*

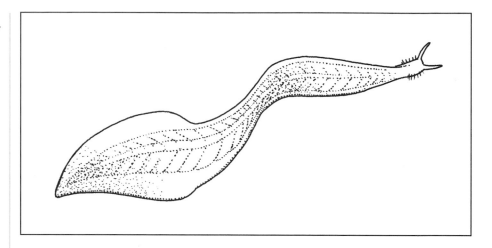

One peculiar animal aptly named *Hallucigenia* (Fig. 55), to honor its peculiar characteristics, was a wormlike creature that propelled itself across the seafloor on what appear to be seven pairs of pointed stilts. However, a more recent interpretation holds that two rows of spines across its back were used for protection. A curious Burgess Shale animal called *Wiwaxia* was another spiny creature about one inch long, possibly related to the modern scaleworm known as a sea mouse. It resembled an undersea porcupine, with large scales and two rows of spikes running along its back apparently for thwarting predators. It fed by scraping off fragments of food from the ocean floor with a rasping organ similar to a horny-toothed tongue.

Figure 55 *The Burgess Shale fauna Hallucigenia is one of the strangest animals preserved in the fossil record.*

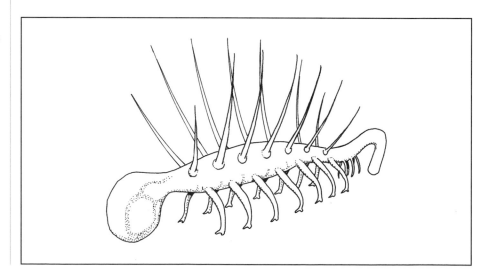

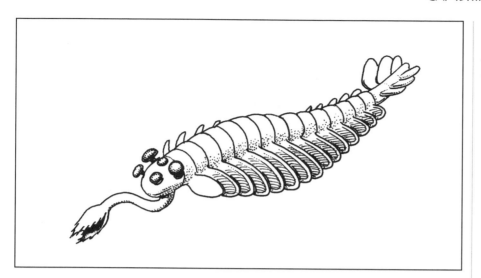

Figure 56 Opabinia
*had a forward-projecting
grasping organ for catching
prey.*

An unusual worm had enormous eyes and prominent fins. A chunky
burrowing carnivorous worm called *Ottoia* was found with shelly animals still
in its gut. When reaching out from the bottom mud, it extended a muscular,
toothed mouth and swallowed its prey whole. An odd creature called *Opabinia*
(Fig. 56) had five eyes arranged across its head, a vertical tail fin to help steer
it along the seafloor, and a grasping organ projected forward possibly used for
catching prey. Its long, hoselike front appendage bestowed upon it the name
"swimming vacuum cleaner."

About half the Burgess Shale fauna consisted of arthropods, with fossil
remains of about 20 different arthropod possibilities that did not survive. One
giant arthropod was as much as 3 feet long. The most interesting of all Burgess
Shale arthropods because of its unusual shape is *Aysheaia,* an animal with a pudgy
body and stubby limbs. An extraordinary arthropod called *Anomalocaris* (Fig. 57),
whose name means "odd shrimp," was possibly the biggest of the Cambrian
predators as well as the oldest-known large predator in the fossil record. This
earliest of monsters reached 6 feet in length and had a mouth surrounded by
spiked plates on the underside of the body. Up to eight rings of teeth reaching
10 inches across were arranged concentrically, leading into the mouth.

Anomalocaris propelled itself by raising and lowering a set of side flaps,
swimming in a manner similar to modern manta rays. It also had a broad tail
with a pair of long trailing spines used for steering and stability. The body was
flanked by a pair of jointed appendages apparently designed for holding and
crushing the armored plates of invertebrates. The animal appeared to be well
equipped for devouring crustaceans and is appropriately dubbed the "terror of
trilobites." Many trilobite fossils are found with rounded chunks bitten out of

Figure 57 Anomalocaris *was a fierce predator of trilobites.*

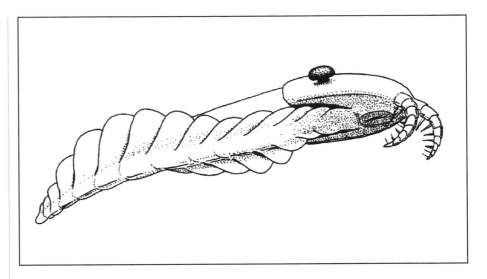

their sides, signifying they had somehow escaped the terrifying grip of *Anomalocaris*. Several trilobite species evolved long spines that might have served as protection against *Anomalocaris* attacks.

Most Burgess Shale species abruptly went extinct at the end of the Cambrian. Although many mass extinctions and proliferations of marine organisms have occurred since, no fundamentally new body styles have appeared during the past 500 million years. After the late Cambrian extinction, only a few archaic forms survived to the middle Devonian. Had they prevailed and produced descendants, Earth would now be graced by an entirely different set of life-forms.

GONDWANA

Near the end of the Precambrian, all landmasses assembled into the supercontinent Rodinia. The continental collision resulted in environmental changes that had a profound influence on the evolution of life. No broad oceans or extreme differences in temperature existed to prevent species from migrating to various parts of the world. Between 630 and 560 million years ago, the supercontinent rifted apart, and four or five continents rapidly drifted away from each other. Most continents huddled near the equator, which explains the presence of warm Cambrian seas. The continental breakup caused sea levels to rise and flood large portions of the land at the beginning of the Cambrian. The extended shoreline might have spurred the explosion of new species, with twice as many phyla living during the Cambrian than before or since.

Many more experimental organisms were in existence than at any other time in Earth history, none of which have any modern counterparts. One example is *Helicoplacus* (Fig. 58), whose body parts were configured in a manner not found in any living organism. It was about 2 inches long and shaped like a spindle covered with a spiraling system of armor plates. It emerged during the transition from the Precambrian to the Cambrian, when more types of body plans arose than at any succeeding period. *Helicoplacus,* as with most species of the early Cambrian, did not leave any living descendants and became extinct about 510 million years ago, just a short 20 million years after it first appeared.

During the Cambrian, continental motions assembled the present continents of South America, Africa, India, Australia, and Antarctica into Gondwana (Fig. 59), named for a geologic province in east-central India. Evidence for Gondwana exists in geologic provinces with similar rock types from the late Precambrian to the early Cambrian. These show matches between Brazil and West Africa; eastern South America, South Africa, West Antarctica, and East Australia; and East Africa, India, East Antarctica, and West Australia. The Pacific margins of South America, Australia, and Antarctica had formed by the beginning of the Cambrian.

East Antarctica was an old Precambrian shield lying to the south of Australia, India, and Africa. A great mountain building episode deformed areas between all pre-Gondwana continents, indicating their collision during this interval. The Transantarctic Mountains arose during the collision of the larger East Antarctica with the geologically younger West Antarctica, composed of volcanic island arcs swept up from the ocean floor.

Much of Gondwana was in the southern polar region from the Cambrian to the Silurian. The present continent of Australia was at the northern

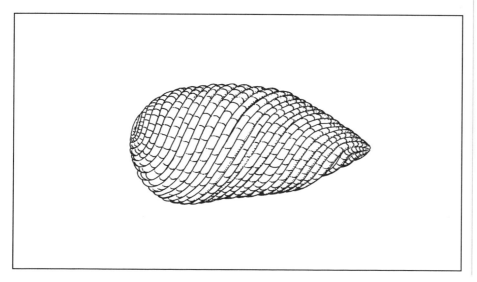

Figure 58 Helicoplacus *was an experimental species whose body parts were arranged in a manner not found in any living creature and became extinct about 510 million years ago after surviving for 20 million years.*

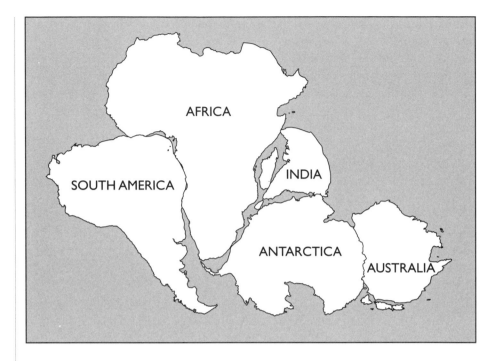

edge of Gondwana and located on the Antarctic Circle. A later collision between North America and Gondwana near the end of the Cambrian about 500 million years ago created an ancestral Appalachian range that continued into western South America long before the Andes formed. North America then broke away from Gondwana and linked with Greenland and Eurasia to form Laurasia about 400 million years ago. Eurasia, the largest modern continent, assembled from about a dozen individual continental plates that welded together at the end of the Proterozoic.

A preponderance of evidence for the existence of Gondwana includes fossilized finds of a mammal-like reptile called *Lystrosaurus* in the Transantarctic Range of Antarctica. They indicate a commonality with southern Africa and India, the only other known sources of *Lystrosaurus* fossils. A fossil of a South American marsupial in Antarctica, which acted as a land bridge between the southern tip of South America and Australia, lends additional support to the existence of Gondwana. Further evidence for Gondwana includes fossils of a reptile called Mesosaurus in eastern South America and South Africa.

After discussing life in the Cambrian, the next chapter takes a look at the early vertebrate life-forms of the Ordovician period.

5

ORDOVICIAN
VERTEBRATES
THE AGE OF SPINAL LIFE-FORMS

This chapter examines the early vertebrate life-forms and the geology of the Ordovician period. The Ordovician, from 500 to 435 million years ago, was named for the ancient Ordovices tribe of Wales, Great Britain. Ordovician marine sediments are recognized on all continents of the Northern Hemisphere, in the Andes Mountains of South America, and in Australia. However, they are absent in Antarctica, Africa, and India. Ordovician terrestrial deposits are not easily recognized because of the lack of fossilized land organisms.

The multitudes of species that exploded onto the scene in the early Cambrian advanced significantly in the warm Ordovician seas (Fig. 60). Corals began building extensive carbonate reefs in the Ordovician. The first fish appeared in the ocean. The existence of freshwater jawless fish suggests that unicellular plants, including red and green algae, inhabited lakes and streams on land.

THE JAWLESS FISH

Beginning about 520 million years ago, the first vertebrates appeared on the scene. They had an internal skeleton made of bone or cartilage, one of life's

Figure 60 *Marine flora and fauna of the late Ordovician.*

(Courtesy Field Museum of Natural History)

most significant advancements. The vertebrate skeleton was light, strong, and flexible, with efficient muscle attachments. The biggest advantage was that the skeleton grew along with the animal. In contrast, invertebrates must shed their external skeletons in order to grow, placing them in great danger to predators. These new internal skeletons enabled the wide dispersal of free-swimming species into a variety of environments.

The most primitive of chordates, which include the vertebrates, was a small, fishlike oddity called amphioxus. Although the animal did not have an actual backbone, it nonetheless is placed in direct line to the vertebrates. The earliest vertebrates lacked jaws, paired fins on either side of the body, or true vertebrae and shared many characteristics of modern lampreys. The origin of the vertebrates led to the evolution of one of the most important novelties— namely the head. It was packed with paired sensory organs, a complex three-part brain, and many other features missing in invertebrates.

The traditional view held that eyes had evolved independently several times in the distant past. However, the discovery of a gene shared by fruit flies, squids, mice, and humans suggest that the eye probably evolved only once in life's evolutionary history. The complex eyes of modern animals probably originally evolved from light-sensitive nerve cells that eventually developed into specialized organs that could focus on a specific object such as prey. Only a minority of the major animal groups have true eyes, however, with 6 out of 30 phyla having complex eyes capable of providing images. However, possessing eyes is such an evolutionary advantage that species with complex eyes comprise 95 percent of the animals on Earth.

Invertebrates, supported by external skeletons, were at a distinct disadvantage in terms of mobility and growth. Many animals such as crustaceans shed their shells as they grew, which often made them vulnerable to predators. One such predator was an extinct giant sea scorpion called eurypterid (Fig. 61), which ranged from the Ordovician to the Permian. It grew to 6 feet long and terrorized creatures on the ocean floor with immense pincers.

The earliest vertebrates were probably wormlike creatures with a prominent rod called a notochord down the back, a system of nerves along the spine, and rows of muscles attached to the backbone arranged in a banded pattern. Rigid structures made of bone or cuticle acted as levers. By using flexible joints, they efficiently translated muscle contractions into organized movements such as rapid lateral flicks of the body to propel the animal through the water. Later, a tail and fins evolved for stabilization. The body became more streamlined and torpedo shaped for speed. With intense competition among the stationary and slow-moving invertebrates, any advancement in mobility was advantageous to the vertebrates.

The oldest known vertebrates were primitive, jawless fish called agnathans (Fig. 62), which first appeared in the early Ordovician about 470 million years ago. Remarkably well-preserved remains of these fish were discovered in the mountains of Bolivia, much of which was inundated by the sea in the Ordovi-

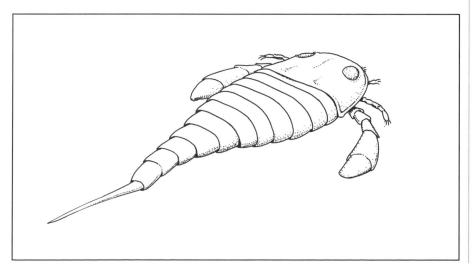

Figure 61 *The extinct eurypterids grew to 6 feet in length.*

Figure 62 *The jawless fish agnathans were the progenitors of fish today.*

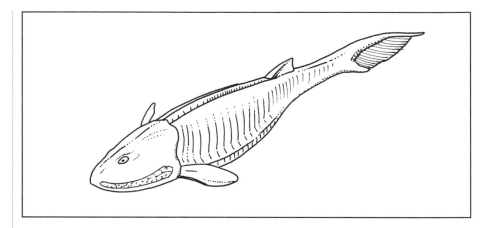

cian. Originally, fossil evidence was scant and fragmentary. Little was known of their appearance or about their evolutionary history. Earlier descriptions dismissed agnathans as a headless, tailless mass of scales and plates or confused the head with the tail, giving it the dubious title of "backward fossil."

The widespread distribution of primitive fish fossils throughout the world suggests a long vertebrate record prior to the Ordovician. The first fish were small mud grubbers and sea squirts that lacked jaws and teeth. The jawless fish whose modern counterparts include lampreys and hagfish had a flexible rod similar to cartilage instead of a bony spine typical of most vertebrates. These ancient fish were probably poor swimmers and avoided deep water. They were generally small, about the size of a minnow, and heavily armored with bony plates protecting the rounded head. The rest of the body was covered with thin scales that ended near a narrow tail. Although well protected from invertebrate predators, the added weight forced the fish to live mostly on the bottom, where they sifted mud for food particles and expelled the waste products through slits on both sides of the throat that later became gills.

Gradually, the protofish acquired jaws and teeth, the bony plates gave way to scales, lateral fins developed on both sides of the lower body for stability, and air bladders maintained buoyancy. Some primitive fish were surprisingly large, up to 18 inches long and 6 inches wide. Thus, for the first time, vertebrates skillfully propelled themselves through the sea. Fish soon became masters of the deep.

FAUNA AND FLORA

Corals are marine coelenterates attached to the ocean floor (Fig. 63). They began constructing extensive carbonate reefs in the Ordovician, building chains of islands and altering the shorelines of the continents. Reef-building corals created the foundations for spectacular underwater edifices that today cover about 750,000 square miles and house about one-quarter of all marine species. The corals diverged into two basic lineages before they developed calcified skeletons, suggesting they might have evolved a reef building ability twice in geologic history. Many corals declined in the late Paleozoic and were replaced by sponges and algae when the seas they inhabited receded.

The bryozoans (Fig. 64), often called moss animals, resemble corals on a smaller scale. However, they are more closely related to brachiopods. They consist of microscopic individuals living in small colonies up to several inches across, giving the ocean floor a mosslike appearance. Bryozoans are retractable creatures, encased in calcareous vaselike structures into which they retreat for safety when threatened. Living species occupy seas at various depths, with certain rare members adapted to life in freshwater.

A single free-moving larval bryozoan establishes a new colony by fixing onto a solid object and growing into many individuals by a process of budding, which is the production of outgrowths. The polyp has a circle of ciliated tentacles, forming a sort of net around the mouth and used for filtering microscopic food floating by. The tentacles rhythmically beat back and forth, producing water currents that aid in capturing food that is digested in a U-shaped gut. Wastes are expelled outside the tentacles just below the mouth.

Fossil bryozoans are common in Paleozoic formations, especially those of the American Midwest and Rocky Mountains. Bryozoan species are iden-

Figure 63 *A collection of coral at Saipan, Marshall Islands.*

(Photo by P. E. Cloud, courtesy USGS)

Figure 64 *The extinct bryozoans were major Paleozoic reef builders.*

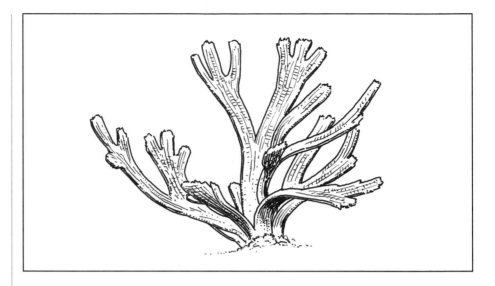

tified by the complex structure of their skeletons, which aids in delineating specific geologic periods. Bryozoans have been very abundant, ranging from the Ordovician to the present. Their fossils are highly useful for making rock correlations. Because of their small size, bryozoans are ideal microfossils for dating oil well cuttings.

Bryozoan fossils are generally found in sedimentary rocks, particularly when covering bedding surfaces. They resemble modern descendants, with some larger groups possibly contributing to Paleozoic reef building, resulting in extensive limestone formations. The fossils are most abundant in limestone and less plentiful in shales and sandstones. Often, a delicate outline of bryozoans can be seen encrusting fossil shells of aquatic animals, stones, and other hard bodies.

Of particular importance to geologists are the ostracods, or mussel shrimp, whose fossils are useful for correlating rocks from the Ordovician onward. Starfish were also common and left fossils in the Ordovician rocks of the central and eastern United States. Their skeletons comprised tiny silicate or calcite plates that were not rigidly joined and therefore usually disintegrated when the animal died, making whole starfish fossils rare. The sea cucumbers have large tube feet modified into tentacles and skeletons comprising isolated plates that are sometimes found as fossils.

As the Ordovician drew to a close, a mass extinction eliminated some 100 families of marine animals 440 million years ago. The climate grew cold, and glaciation reached its peak. Ice sheets radiated outward from a center in North Africa, which hovered directly over the South Pole. Most of the victims were tropical species sensitive to fluctuations in the environment. Among those that went extinct were many trilobite species. Before the extinction, trilobites accounted for about two-thirds of all species but only a third thereafter. One dominant group of trilobites called the Ibex fauna vanished at the end-Ordovician mass extinction possibly due to widespread glaciation. They were succeeded by the Whiterock fauna, which tripled their genera and sailed through the extinction practically unscathed.

Graptolites (Fig. 65) were colonies of cupped organisms resembling a conglomeration of floating stems and leaves. Certain groups went extinct at the end of the Ordovician. Graptolites were thought to have suffered total extinction in the late Carboniferous about 300 million years ago. However, the discovery of living pterobranchs, possible modern counterparts of graptolites, suggest these might be living fossils.

Near the end of the Ordovician 450 million years ago, the concentration of atmospheric oxygen generated sufficient levels of ozone in the upper stratosphere to shield Earth from the Sun's deadly untraviolet rays. Thus, for the first time, plants began to come ashore to populate the land. When the early plants first left the oceans and lakes for a home on dry land, they were greeted by a harsh environment. Ultraviolet radiation, desert conditions, and lack of nutrition made life difficult. First to greet the land plants were soil bacteria that churned sediments into lumpy brown mounds. The presence of these bacteria helped speed weathering processes, without which hot bare rock would have covered most of the landscape and land plants would have had little success gaining a roothold.

Cyanobacteria, formerly called blue-green algae, might have been preparing the soil for the land invasion as early as 3 billion years ago. Ancient cyanobacteria, which were resistant to high levels of ultraviolet radiation, first lived in shallow tide pools, from which they eventually colonized the continents. They might have improved the terrestrial climate for a life out of water by drawing down atmospheric carbon dioxide, thereby countering the greenhouse effect, which prevents thermal energy from escaping the planet. The soil

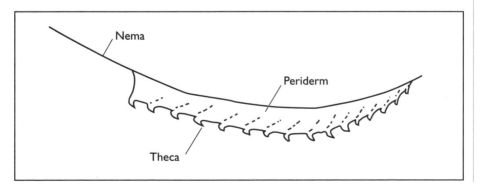

Figure 65 Graptolites appeared to have gone extinct in the late Carboniferous.

Figure 66 *Rafts of blue-green algae in a pool in Indian Canyon, Duchesne County, Utah.*

(Photo by W. H. Bradley, courtesy USGS)

bacteria helped resist erosion by binding the sediment grains together and soaking up rainwater. Bacteria also provided nutrients for the early land plants.

Plant fossils of Ordovician age appear to be almost entirely composed of algae similar to present-day algal mats found on seashores and the bottoms of ponds (Fig. 66). Some marine algae lived in the intertidal zones and could withstand only short periods out of the sea due to the risk of dehydration. Even after developing protective measures to help the organisms survive out of water for longer periods, they still depended on the sea for reproduction.

Lichens, which are a symbiotic relationship between algae and fungi whereby the two mutually live off each other, probably took the first tentative steps onto dry land. Following the lichens were mosses and liverworts. Fungi also had a symbiotic relationship with the roots of plants when they first evolved, aiding vegetation with the uptake of nutrients and receiving carbohydrates in return. Bacteria also played an important role in the fixation of nitrogen, an abundant soil gas and an important nutrient for plant life. In the nitrogen cycle, denitrifying bacteria convert the dissolved nitrate back into elemental nitrogen. Otherwise, all nitrogen in the atmosphere would have long ago disappeared.

THE ORDOVICIAN ICE AGE

Species that rapidly evolved during the early Cambrian advanced significantly in the warm Ordovician seas. The warming was largely because the atmosphere held as much as 16 times today's carbon dioxide content, enough to heat the climate to tropical levels even though the Sun was 4 percent dimmer than at present. The average global temperature was about 18 degrees Celsius, some 8 degrees hotter than today. Corals, which require warm waters, began building extensive carbonate reefs. In addition, the first fish appeared in the ocean. The existence of freshwater jawless fish on the continents suggests that lakes and streams were inhabited by red and green algae.

Plants began to invade the land and extend to all parts of the world during the late Ordovician about 450 million years ago. The early land plants absorbed large quantities of atmospheric carbon dioxide. Rapid burial under anaerobic conditions deposited the organic carbon into the geologic column, where it converted into coal. Plants also aided the weathering process by leaching minerals from the rocks. Carbonate rocks, such as limestone deposited by shelly organisms from the Cambrian onward (Fig. 67), locked up massive amounts of carbon dioxide.

The withdrawal of substantial amounts of carbon dioxide from the atmosphere weakened the greenhouse effect. The resulting climate cooling initiated in large part by the plant invasion spawned a major ice age at the end of the Ordovician about 440 million years ago. At the time, the southern edge of Gondwana was just over the South Pole, where an ice sheet grew to about 80 percent the size of present-day Antarctica. The glaciations of the late Ordovician and the glacial epochs of the middle and late Carboniferous about 330 million and 290 million years ago might have been influenced by a reduction of atmospheric carbon dioxide to roughly one-quarter of its present value.

Atmospheric scientists have amassed information on global geochemical cycles to ascertain the cause of such a radical change in the carbon dioxide content of the atmosphere. Data from deep-sea cores showed that carbon dioxide variations preceded changes in the extent of the more recent ice ages, suggesting that earlier glacial epochs might have been similarly affected. The variations of carbon dioxide levels might not be the sole cause of glaciation. However, when combined with other processes, such as variations in Earth's orbital motions or a drop in solar radiation, they could become a strong influence.

Continental movements might also be responsible for the late Ordovician glaciation. Magnetic orientations in rocks from many parts of the world indicate the positions of continents relative to the magnetic poles at various times in Earth history. Paleomagnetic studies in Africa revealed very curious findings, however. They placed North Africa directly over the South Pole during the Ordovician, which led to worldwide glaciation.

Additional evidence for such widespread glaciation came from another surprising location—the middle of the Sahara Desert. Geologists exploring for petroleum in the region stumbled upon a series of giant grooves cut into the underlying strata by glaciers. Rocks embedded at the base of glaciers scoured the landscape as the ice sheets moved back and forth. Other collaborating evidence suggests that thick sheets of ice blanketed the Sahara and included erratic boulders placed by moving ice and eskers, which are sinuous sand deposits from glacial outwash streams.

A major mountain building episode from the Cambrian to the Ordovician deformed areas between all continents comprising the southern supercontinent Gondwana, indicating their collision during this interval. The

Figure 67 Intensely folded Cambrian carbonate rocks of Scapegoat Mountain, Lewis and Clark County, Montana.

(Photo by M. R. Mudge, courtesy USGS)

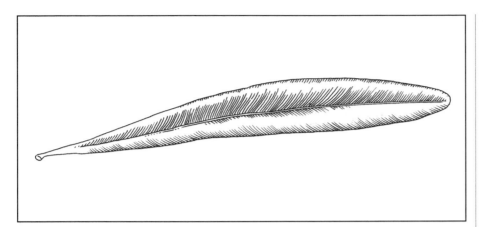

Figure 68 *Fossil* Glossopteris *leaves helped prove the theory of continent drift.*

middle Paleozoic fern *Glossopteris* (Fig. 68), named from the Greek word meaning "featherlike," and whose fossil leaf impressions actually look like feathers, is found in coal beds on the southern continents and India. However, the plant is suspiciously absent on the northern continents. This suggests the existence of two large continents, one located in the Southern Hemisphere and another in the Northern Hemisphere, separated by a large open sea. Matches between mountains in Canada, Scotland, and Norway indicated their assembly into the northern supercontinent Laurasia during this time.

At the end of the Ordovician, glaciation reached its peak. Ice sheets radiated outward from a center in North Africa. Around 430 million years ago, the ice sheets largely disappeared. As Gondwana continued drifting southward, the farther south it went, the smaller the ice sheets became. When the center of the continent neared the South Pole, the winters in the interior became colder. Yet the land warmed sufficiently during the summer to melt the ice. Meanwhile, the southern glaciated edge of Gondwana moved northward into warmer seas, and the glaciers soon departed.

THE IAPETUS SEA

During the late Precambrian and early Cambrian, a proto–Atlantic Ocean called the Iapetus opened between divided continents, forming extensive Cambrian inland seas. The inundation submerged most of the surrounding ancient North American continent called Laurentia and the ancient European continent called Baltica. The Iapetus Sea was similar in size to the North Atlantic and occupied the same general location about 500 million years ago (Fig. 69). A continuous coastline running from Georgia to Newfoundland

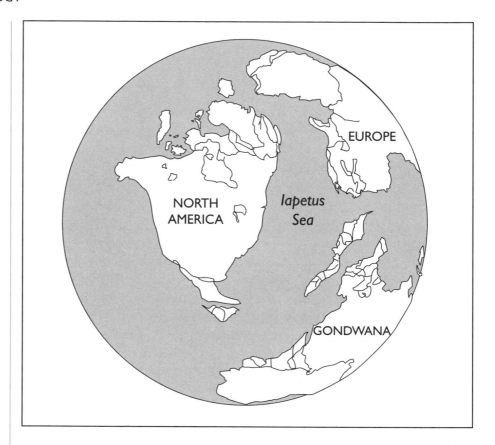

between about 570 and 480 million years ago suggests that this ancient sea-coast faced a wide, deep ocean.

The Iapetus stretched at least 1,000 miles across from east to west and bordered a much larger body of water to the south. It was dotted with volcanic islands and resembled the present-day Pacific Ocean between Southeast Asia and Australia. The shallow waters of the nearshore environment of this ancient sea from the Cambrian to the mid-Ordovician, about 460 million years ago, contained abundant invertebrates, including trilobites, which accounted for about 70 percent of all species. Eventually, the trilobites faded, while mollusks and other invertebrates expanded throughout the seas.

The closing of this ancient ocean basin in the Ordovician as Baltica approached Laurentia signaled the formation of Laurasia. Large-scale mountain building followed the closing of the Iapetus when the continents flanking the sea collided, pushing up mountains in northern Europe and North America, including those that evolved into the Appalachians. The spate of mountain building might have triggered a burst of species diversity. The largest of the increases was the Ordovician radiation of marine species around 450 million years ago.

The foreland basins filled with thick sediments eroded from nearby mountains. Erosion from the mountains might have pumped nutrients into the sea, fueling booms in marine plankton, thereby increasing the food supply for higher creatures. Therefore, the number of genera of mollusks, brachiopods, and trilobites (Fig. 70) dramatically increased, because organisms with abundant food are more likely to thrive and diversify into different species.

Island arcs lying between the two colliding landmasses were scooped up and plastered against continental edges as the two continents collided. The oceanic crustal plate carrying the islands dived under Baltica in a process known as subduction. The subduction rafted the islands into collision with the

Figure 70 Fossil brachiopods and trilobites from the Bonanza King Formation, Trail Canyon, Death Valley National Monument, Inyo County, California.

(Photo by C. B. Hunt, courtesy USGS)

Figure 71 *Steeply dipping Paleozoic rocks of the Brooks Range, Anaktuvuk district, Northern Alaska.*

(Photo by J. C. Reed, courtesy U.S. Navy and USGS)

continent and deposited the formerly submerged rocks onto the present west coast of Norway. Slices of land called terranes residing in western Europe migrated into the Iapetus from ancient Africa. In the same manner, slivers of crust from Asia traveled across the ancient Pacific Ocean called the Panthalassa, Greek for "universal sea," to form much of western North America.

A large portion of the Alaskan panhandle, known as the Alexander Terrane, began its existence as part of eastern Australia some 500 million years ago. About 375 million years ago, it broke free from Australia, traversed the proto–Pacific Ocean, stopped briefly at the coast of Peru, slid past California, and rammed into the upper North American continent around 100 million years ago. The entire state of Alaska is an agglomeration of terranes that were pieces of ancient oceanic crust. Terranes are well exposed in the Brooks Range (Fig. 71), a major east-west–trending mountain belt that makes up the spine of northern Alaska. Basaltic seamounts that accreted to the margin of North America traveled halfway across the ocean that preceded the Pacific.

Terranes (Fig. 72) are fault-bounded blocks. They range in size from small crustal fragments to subcontinents, with geologic histories markedly different from those of neighboring blocks and of adjoining continental masses.

(Note: Do not confuse the term *terrane* with the word *terrain,* which means landform.) Terranes are usually bounded by faults and are distinct from their geologic surroundings. The boundary between two or more terranes is called a suture zone. The composition of terranes generally resembles that of an oceanic island or plateau. Others are composed of a consolidated conglomerate of pebbles, sand, and silt that accumulated in an ocean basin between colliding crustal fragments.

Terranes, which are as much as 1 billion years old, are dated by analyzing entrained fossil radiolarians (Fig. 73), marine protozoans that lived in deep water and were abundant from about 500 million to 160 million years ago. Different species also defined specific regions of the ocean where the terranes originated. Many terranes traveled great distances before finally adhering to a continental margin. Some North American terranes have a western Pacific origin and were displaced thousands of miles to the east.

From the Cambrian to the end of the Paleozoic, the western edge of North America ended near present-day Salt Lake City. Over the last 200 million years, North America has expanded by more than 25 percent during a major pulse of crustal growth. Much of western North America was assembled from oceanic island arcs and other crustal debris skimmed off the Pacific plate as the North American plate headed westward after the breakup of the supercontinent Pangaea.

Terranes exist in a variety of shapes and sizes, ranging from small slivers to subcontinents such as India, itself a single great terrane. Most terranes are elongated bodies that deform when colliding with a continent. Terranes cre-

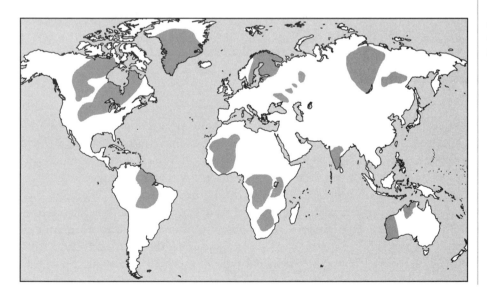

Figure 72 *Distribution of 2-billion-year-old terranes.*

Figure 73 *Upper Devonian radiolarians from the Kandik Basin, Yukon Region, Alaska.*

(Photo by D. L. Jones, courtesy USGS)

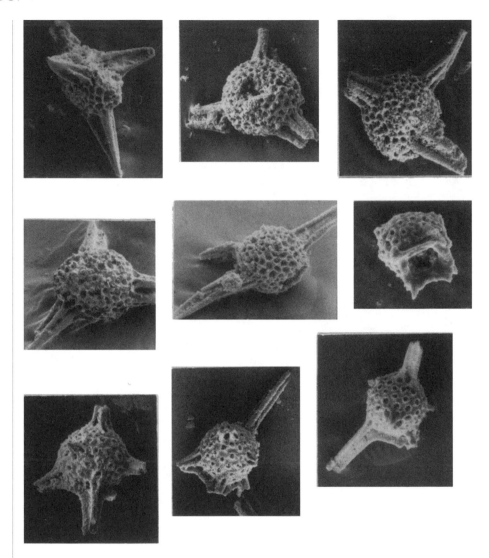

ated on an oceanic plate retain their shapes until they collide and accrete to a continent. They are then subjected to crustal movements that modify their overall dimensions. The assemblage of terranes in China is being stretched and displaced in an east–west direction due to the continuing squeeze India is exerting onto southern Asia as it raises the Himalayas.

Granulite terranes are high-temperature metamorphic belts formed in the deeper parts of continental rifts. They also comprise the roots of mountain belts created by continental collision, such as the Alps and Himalayas. North of the Himalayas is a belt of ophiolites, which marks the boundary between the sutured continents. Terrane boundaries are commonly marked by ophiolite belts, consisting of marine sedimentary rocks, pillow basalts, sheeted dike complexes, gabbros, and peridotites.

Terranes also played a major role in the creation of mountain chains along convergent continental margins. For example, the Andes appeared to have been raised by the accretion of oceanic plateaus along the continental margin of South America. Along the mountain ranges in western North America, the terranes are elongated bodies due to the slicing of the crust by a network of northwest-trending faults. One of these is the San Andreas Fault in California, which has undergone some 200 miles of displacement in the last 25 million years.

Around 500 million years ago, North America was a lost continent. South America, Africa, Australia, Antarctica, and India had assembled into the supercontinent Gondwana. However, North America and a few smaller continental fragments were drifting freely on their own. At this time, North America was situated a few thousand miles off the western coast of South America, placing it on the western side of Gondwana. About 750 million years ago, North America lay at the core of an earlier supercontinent called Rodinia, when Australia and Antarctica bordered the west coast of the North American continent.

North and South America apparently abutted one another at the beginning of the Ordovician (Fig. 74), placing what would be present-day Washington, D.C., close to Lima, Peru. A limestone formation in Argentina contains

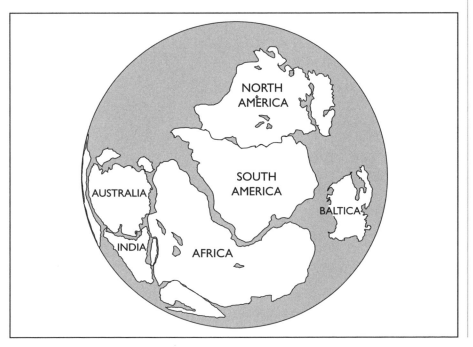

Figure 74 North America (top) *abutted South America* (center) *during the early Ordovician.*

a distinct trilobite species typical of North America but not of South America. The fossil evidence suggests that the two continents collided about 450 million years ago, creating an ancestral Appalachian range along eastern North America and western South America long before the present Andes formed. Later, the continents rifted apart, transferring a slice of land containing trilobite fauna from North America to South America.

THE CALEDONIAN OROGENY

As Laurentia approached Baltica at the end of the Silurian about 400 million years ago, they closed off the Iapetus Sea some 200 million years before the modern Atlantic began to open. When the continents collided, they crumpled the crust and forced up mountain ranges at the point of impact. The sutures joining the landmasses are preserved as eroded cores of ancient mountains called orogens. Paleozoic continental collisions raised huge masses of rocks into several folded mountain belts throughout the world. A major mountain building episode from the Cambrian to the middle Ordovician deformed areas between all continents comprising Gondwana, indicating their collision during this interval.

Matching geologic provinces exist between South America, Africa, Antarctica, Australia, and India. The Cape Mountains in South Africa have counterparts with the Sierra Mountains south of Buenos Aires in Argentina. Matches also exist between mountains in Canada, Scotland, and Norway. During this time, much of Gondwana was in the southern polar region, where glaciers expanded across the continent, leading to an Ordovician ice age.

The closing of the Iapetus Sea from the middle Ordovician to the Devonian as Laurentia approached Baltica resulted in the great Caledonian orogeny (Fig. 75), or mountain building episode. This orogenic activity formed a mountain belt that extended from southern Wales, spanned Scotland, and ran through Scandinavia and Greenland, possibly including today's extreme northwest Africa as well. In North America, this orogeny built a mountain belt that extended from Alabama through Newfoundland and reached as far west as Wisconsin and Iowa. Vermont still preserves the roots of these ancient mountains, which were shoved upward between about 470 and 400 million years ago but have since planed off by erosion.

The middle Ordovician Taconian orogeny named for the Taconic Range of eastern New York State culminated in a chain of folded mountains that extended from Newfoundland through the Canadian Maritime Provinces and New England, reaching as far south as Alabama. During the Taconian disturbance, extensive volcanic activity occurred in Quebec and Newfoundland and from Alabama to New York, extending as far west as Wisconsin and Iowa.

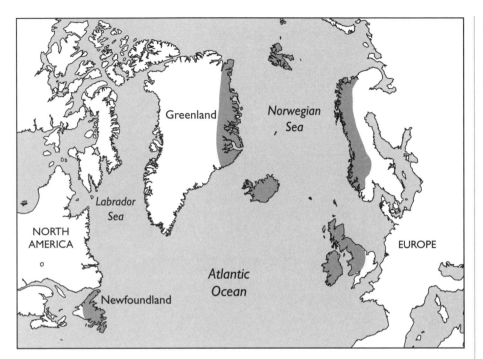

An inland sea flooded the continent in the middle Ordovician and reached a maximum in the late Ordovician. It partially withdrew in response to a flood of sediments eroded from the Taconian mountain belts. One of these sedimentary deposits is the widespread Ordovician St. Peter Sandstone of the central United States. It is composed of well-sorted, nearly pure quartz beach sands that are ideal for manufacturing glass.

After covering life and geology in the Ordovician, the next chapter explores the first life-forms to leave the sea and settle on dry land during the Silurian period.

6

SILURIAN PLANTS
THE AGE OF TERRESTRIAL FLORAS

This chapter examines the plant life and geology of the Silurian period. The Silurian, from 440 to 400 million years ago, was named for the Silures, an ancient Celtic tribe of Wales, Great Britain. Many of today's mountain ranges were uplifted by middle Paleozoic continental collisions. The clashing of North America with Eurasia to create Laurasia closed off the Iapetus Sea during the Silurian. The collision formed mountain belts on the margins of continents surrounding the ancient sea, producing intensely folded rocks (Fig. 76).

Evidence of widespread reef building by Silurian corals indicates the existence of warm, shallow seas with little temperature variation. The tabulate corals, comprising closely packed polygonal or rounded calcite cups or thecae, were another important group of reef builders. The rugose, or horn corals, so named because of their typical hornlike shapes, were particularly abundant in the Silurian. They were the major reef builders of the late Paleozoic, finally going extinct in the early Triassic. The higher land plants were firmly established on the previously barren continents. Eventually, creatures crawled out of the sea to dine on them.

THE AGE OF SEAWEED

Like animals, plants did not appear in the fossil record as complex organisms until the late Precambrian or early Cambrian. The distinction between plants and animals is somewhat blurred in the geologic record of the obscure past, when they shared many common characteristics. From humble beginnings as simple algae, single-celled plants probably colonized for similar reasons that unicellular animals grouped together, such as mutual support, division of labor, and protection. However, complex marine plants did not appear in the fossil record until the Cambrian, after which they evolved rapidly.

Although the Cambrian has been referred to as "the age of seaweed," the geologic record does not support this contention with strong fossil evidence. Well-preserved multicellular algae and a variety of fossil spores discovered in late Precambrian and Cambrian sediments suggest the existence of complex sea plants. However, no other significant remains have been found. Even as late

Figure 76 *Folded sandstones and shales near the base of the upper Silurian near Hancock, Washington County, Maryland.*

(Photo by C. D. Walcott, courtesy USGS)

Figure 77 *Algal mounds rising toward the surface from the surrounding limestone bottom of a barrier reef in Saipan, Mariana Islands.*

(Photo by P. T. Cloud, courtesy USGS)

as the Ordovician, plant fossils appeared to be composed almost entirely of algae, which probably formed stromatolite mounds and algal mats similar to those on seashores today (Fig. 77).

The early seaweeds were soft and nonresistant. Generally, they did not fossilize well. A seaweedlike plant grew half submerged in estuaries and rivers. However, for plants to be truly shore bound, they had to reproduce entirely out of water. The first land plants achieved this function with sacs of spores attached to the ends of simple branches. When the spores matured, they were cast into the air and carried by the wind to suitable sites where they could grow into new plants.

The first complex plants lived in shallow waters just below the surface probably as a protection against high levels of solar ultraviolet radiation. When the atmospheric oxygen content rose to near present-day levels, the upper stratospheric ozone layer screened out the deadly ultraviolet rays, enabling life to flourish on Earth's surface. Soon after plants crept ashore, the land was sprawling with lush forests.

Before the invasion of the true plants, a slimy coating of photosynthesizing cyanobacteria, or blue-green algae, might have inhabited the land. A cover of algae accelerated the weathering of rock and the formation of soils and nutrients required for advanced plant life. Prior to the emergence of land plants, microbial soils were making Earth more hospitable for life out of water. The microorganisms probably formed a dark, knobby soil, resembling lumpy mounds of brown sugar spread over the landscape. In this manner, for about half a billion years, simple plants paved the way for more advanced vegetation.

Prior to the arrival of terrestrial microbes, the continents were much too hot to support complex life-forms. The early organisms played an important role in cooling the land surface by drawing down the atmosphere's surplus carbon dioxide for use in photosynthesis. The loss of this potent greenhouse gas cooled the climate and allowed higher forms of life to populate the continents. The microbes also aided in the weathering of rock into soil, helped to prevent soil erosion, and provided the nutrients more advanced plants needed for survival.

The first land plants comprised algae and seaweedlike plants. They resided just below the sea surface in the shallow waters of the intertidal zones (Fig. 78). Primitive forms of lichen and moss lived on exposed surfaces. They were followed by tiny fernlike plants called psilophytes, or whisk ferns, the predecessors of trees. They were among the first plants to live onshore. These simple plants lived semisubmerged in the intertidal zones, lacked root systems and leaves, and reproduced by casting spores into the sea for dispersal. The most complex land plants grew less than an inch tall and resembled an outdoor carpet covering the landscape.

By the late Silurian, all major plant phyla were in existence. Except for simple algae and bacteria, the early land plants diverged into two major groups. The bryophytes, including mosses and liverworts, were the first plant phylum to become well established on land. They have stems and simple leaves. However, they lack true roots or vascular tissues to conduct water to the higher extremities and therefore are required to live in moist environments. They reproduce by spores that are carried by the wind for wide distribution. The earliest species occupied freshwater lakes in the late Precambrian.

The pteridophytes, or ferns, were the first plant phylum to develop true roots, stems, and leaves. Some present-day tropical ferns grow to tree size as they did in the geologic past. The whisk ferns, which appeared at the end of the Silurian and became extinct near the end of the Devonian, probably gave rise to the first club mosses, horsetails, and true ferns. The true ferns are the largest group both living and extinct and contributed substantially to

Figure 78 *Evolution of plants from the ocean to the land:* (1) *fully submerged;* (2) *semisubmerged;* (3) *fully terrestrial.*

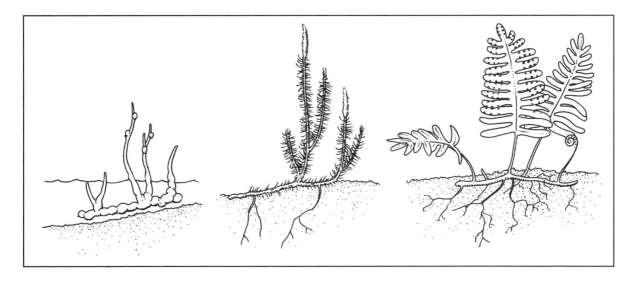

Carboniferous coal deposits. Most reproduced with spores, but the extinct seed ferns bore seeds.

The spermatophytes are the higher plants that produce seeds and include the gymnosperms and angiosperms. Gymnosperms are conifers that bare seeds on exposed scales or cones. They ranged from the Carboniferous to the present and covered vast areas in harsh climates. Angiosperms are flowering plants, whose seeds often develop in a fruit. They originated in the Cretaceous and range in size from grasses to huge trees. The higher plants are represented by about 270,000 modern species.

The next major evolutionary stage was the development of a vascular stem using channels to conduct water from a swamp or from the moist ground nearby to the plant's extremities. The strengthening of the stems enabled vascular plants to grow tall. The early club mosses, ferns, and horsetails were the first plants to make use of this system. When roots evolved, plants could survive entirely on dry land by drawing water into their stalks from soil moistened by rain. By the end of the Silurian 400 million years ago, 2-inch-long plants had water-conducting vessels that allowed them live totally out of water.

The lycopods, which included the club mosses and scale trees (Fig. 79), were the first plants to develop true roots and leaves. The branches were arranged in a spiral shape, and leaves were generally small. Spores were attached to modified leaves that became primitive cones. The scale trees, so named because the scars on their trunks resembled large fish scales, grew upward of 100 feet or more high and eventually became one of the dominant trees of the Paleozoic forests.

Many plants were still quite primitive and restricted to wet habitats. They had no roots or leaves and resembled branching sticks creeping along the ground. More advanced plants grew as high as 10 feet tall and had substantial root systems that burrowed into the ground as much as 3 feet deep. The deep-rooted plants lived along streams that seasonally dried up, requiring them to tolerate prolonged arid conditions. Since the environment was so harsh, having large roots helped the plants withstand such severe conditions. The deep roots helped control erosion, allowing the buildup of fertile soils on the land, which encouraged additional plant growth.

The burgeoning vegetation also produced a substantial drop in atmospheric carbon dioxide concentrations, changing a steaming greenhouse climate to one more hospitable for the first animals to populate the land. The plants shared the land with a variety of arthropods. They were joined by amphibious fish, which began to make short forays onto the beach to prey on abundant crustaceans and insects. Freshwater invertebrates and fish began to inhabit the lakes and streams.

During their first 50 million years on dry land, plants displayed increased diversity and complexity, including root systems, leaves, and reproductive organs employing seeds instead of spores. When true leaves evolved, plants developed a

Figure 79 *The scale tree was one of many early trees in the lush Carboniferous forest.*

variety of branching patterns to expose them to as much sunlight as possible to maximize photosynthesis. Competition for light was of primary importance to the evolution of land plants. Those that developed the most efficient branching patterns gathered the most light and, therefore, were the most successful.

As plants grew larger, they progressed from random branching to tiers of branches to achieve greater efficiency with a minimum of self-shading similar to present-day pines. This added weight increased the mechanical stress on the plant, requiring stronger branches to prevent limbs from breaking during storms. These innovations gave rise to the preponderance of floras in existence today.

THE REEF BUILDERS

Silurian marine invertebrates (Fig. 80) were intermediate in evolution between Ordovician and Devonian lines. The Silurian is paleontology's "dark

Figure 80 *Marine flora and fauna of the middle Silurian.*

(Courtesy Field Museum of Natural History)

age" in the sense that little is known about the soft bodies of animals that inhabited the seas. Soft-bodied marine animals living in the Silurian some 430 million years ago included a host of worms and bizarre arthropods. Among the odd animals were a half-inch-long bristly worm and a tiny shrimplike creature with a tentacled head, a segmented body, and a triangular tail. More enigmatic yet was a half-inch-long creature that might be an unknown relative of a group of small stubby-legged worms called lobopods. A large variety of wormlike creatures had their beginning in the Cambrian and apparently evolved into higher forms of animal life.

Coral reef formation during the Silurian was widespread, indicating the presence of warm shallow seas with little seasonal temperature variation. Corals began constructing extensive reefs in the Ordovician, forming barrier islands and island chains. They also built atolls atop extinct submerged volcanoes. As the volcanoes subsided beneath the sea, the rate of coral growth matched the rate of volcanic subsidence. This kept the corals at a constant shallow depth for photosynthesis.

Coral reefs are among the oldest ecosystems on Earth and important land builders. They form chains of islands and alter continental shorelines. The major structural feature of the reef is the coral rampart, which reaches almost to the water's surface. It consists of large rounded coral heads and a variety of branching corals. The fore reef is seaward of the reef crest, where coral blankets nearly the entire seafloor. In deeper waters, many corals grow in flat, thin sheets to maximize their light-gathering area.

In other parts of the reef, the corals form large buttresses separated by narrow, sandy channels composed of calcareous debris from dead corals, calcareous algae, and other organisms living on the coral. The channels resemble narrow, winding canyons with vertical walls of solid coral. They dissipate wave energy and allow the free flow of sediments, which prevents the coral from choking on the debris. Below the fore reef is a coral terrace, followed by a sandy slope with isolated coral pinnacles, then another terrace, and finally a near vertical drop into the dark abyss.

The coral polyp (Fig. 81) is a soft-bodied animal crowned by a ring of tentacles surrounding a mouthlike opening. The tentacles are tipped with poisonous stingers to attack prey swimming nearby. The polyp lives in a skeletal cup or tube called a theca composed of calcium carbonate. Polyps extend their tentacles to feed at night and withdraw into their thecae during the day or at low tide to keep from drying out in the Sun. Because the algae within the polyp's body require sunlight for photosynthesis, corals are restricted to warm, shallow water generally less than 100 feet deep.

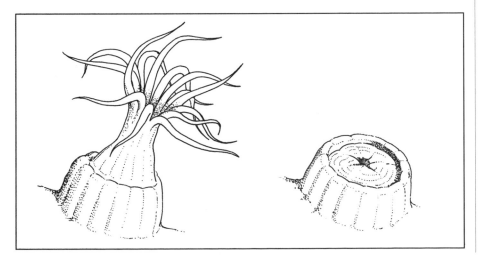

Figure 81 *Coral polyps seek protection in carbonate cups from predators and during low tide.*

A large variety of corals are well represented in the fossil record and resemble many of their modern counterparts. The tabulate corals, which became extinct at the end of the Paleozoic, consisted of closely packed polygonal or rounded thecae, some with pores covering the walls of the thecae. The rugose, or horn corals, named for their hornlike shape, were particularly abundant in the Silurian. They became the major reef builders of the late Paleozoic, finally becoming extinct in the early Triassic. The hexacorals, with thecae separated by six sepia, or walls, ranged from the Triassic to today and were the major reef builders of the Mesozoic and Cenozoic eras. The tetracorals (Fig. 82), whose sepia were arranged in groups of four, were another extinct group of reef-building corals. The rudists (Fig. 83) were mollusks that displaced corals as the dominant reef builders in the Cretaceous and became extinct at the end of the period.

In the geologic past, massive coral structures turned into some of the greatest limestone deposits on Earth. Corals constructed barrier reefs and atolls. They also played a major role in changing the face of the planet. Coral reefs contain abundant organic material. Many ancient reefs are composed largely of a carbonate mud with the skeletal remains of a variety of other species literally "floating" in the sediment, producing some of the finest fossil specimens.

Tropical plant and animal communities thrived on the reefs due to the coral's ability to build massive wave-resistant structures. Unfortunately, these were the same species that suffered several episodes of extinction due to their narrow range of living conditions. The extinctions hit hardest those organisms anchored to the ocean floor or unable to migrate out of the region because of physical and biological barriers. Despite the excellent climate and the extreme success of the trilobites in the lower Paleozoic, they began to decline

Figure 82 *The extinct tetracorals were major reef builders.*

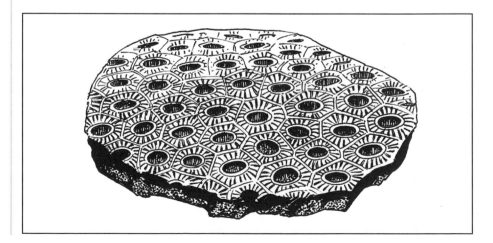

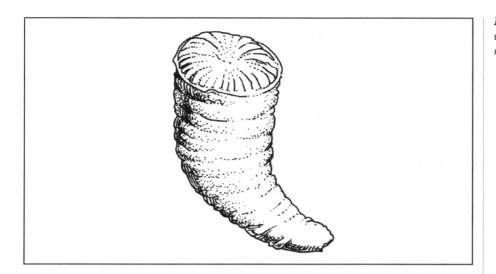

Figure 83 *The rudists were major reef-building mollusks.*

rapidly in the Silurian. Numerous species succumbed to extinction at the end of the period.

The echinoderms with fivefold and bilateral symmetries and exoskeletons composed of numerous calcite plates were among the most prolific animals of the Silurian seas. The most successful of the Silurian echinoderms were the crinoids, commonly called sea lilies because they resemble flowers atop stalks anchored to the seabed. Some crinoids were also planktonic or free-swimming types (Fig. 84). They became the dominant echinoderms of the middle and upper Paleozoic, with many species still in existence.

The long stalks of the crinoids grew upward of 10 feet in length or longer in good conditions. They were composed of perhaps 100 or more calcite disks called columnals and anchored to the ocean floor by a rootlike appendage. A cup called a calyx perched on top of the stalk and housed the digestive and reproductive organs. The animal strained food particles from passing water currents with five feathery arms that extended from the calyx, giving the crinoid a flowerlike appearance. The extinct Paleozoic crinoids and their blastoid relatives, whose calyx resembled a rosebud, made excellent fossils, especially the stalks, which on weathered limestone outcrops often looked like long strings of beads.

The echinoids are a class of echinoderms that include sea urchins, heart urchins, and sand dollars. They have exoskeletons composed of limy plates that are characteristically spiny, spherical, or radially symmetrical. Some more advanced forms were elongated and bilaterally symmetrical about a central point. The sea urchins lived mostly among rocks encrusted with algae, upon which they fed. Unfortunately, such an environment was not conducive to fossilization. Likewise, the familiar sand dollars that occasionally washed up on beaches are rare in the fossil record.

Figure 84 *Specimen of a microcrinoid from the lower member of the Lawson Limestone in the Florida Peninsula, Lafayette County, Florida.*

(Photo by P. L. Applin, courtesy USGS)

THE LAND INVASION

The colonization of the continents is one of the most important steps in evolutionary history. Strangely, after having been in existence for more than three-quarters of Earth history, life finally conquered the land some 450 million years ago. Part of the reason for such a delay might be the lack of oxygen during the previous era needed to form an effective ozone screen. The ozone layer filters out harmful ultraviolet rays from the Sun. The strong ultraviolet radiation probably kept life in the protective waters of the oceans until the concentration of ozone was sufficient to make conditions safe to venture out onto dry land.

The first invertebrates to crawl out of the sea and populate the continents were probably crustaceans. These ancient arthropods emerged from the ocean soon after plants began to colonize the land. The oldest known land-adapted animals were centipedes and tiny spiderlike arachnids about the size of a flea and found in Silurian rocks some 415 million years old. The arachnids are air-breathing crustaceans and include spiders, scorpions, daddy long-legs, ticks, and mites. The early terrestrial communities consisted of small plant-eating arthropods that served as prey for the arachnids, which were predatory animals.

The early crustaceans were segmented creatures, ancestors of today's millipedes, and walked on perhaps 100 pairs of legs. At first, they remained near shore, eventually moving farther inland along with the mosses and lichens. Because they had the land to themselves, no competitors, and an abundant food supply, some species evolved into the first terrestrial giants—growing up to 6 feet long. The crustaceans became easy prey for the descendants of giant sea scorpions called eurypterids when they eventually came ashore.

The advent of forests, where leaves and other edible parts grew beyond easy reach from the ground, posed new problems for the ancestors of the insects. Climbing up tall tree trunks to feed on stems and leaves was probably less dangerous than the treacherous journey back down. The return to the ground would have been much easier if they simply jumped or glided through the air on primitive winglike structures. These appendages probably originated as a means to regulate the insect's body temperature. By natural selection, they developed into flapping wings. They worked well for launching insects to the treetops and for escaping predators when the vertebrates finally came to shore.

Insects are by far the largest living group of arthropods. They can easily claim the title of the world's most prosperous as well as the most populous creatures. Insects and plants have fought with each other for more than 300 million years. The fiercest battles have been played out in the tropics, where hordes of hungry pests attack vegetation, which in turn defend themselves with poisons.

Ever since animals left the sea and took up a lifestyle on dry land, insects and other arthropods have ruled the planet. Insects have three pairs of legs and typically two pairs of wings on the thorax, or midsection. In most cases, an insect's body is covered with an exoskeleton made of chitin, similar to cellulose. To achieve flight, insects had to be lightweight. As a result, their delicate bodies did not fossilize well unless trapped in tree sap, which became hard amber, allowing the insects to withstand the rigors of time. In some groups, the exoskeleton was composed of calcite or calcium phosphate, which enhanced these insects' chances of fossilization.

LAURASIA

During the Silurian, all northern continents collided to form Laurasia, which included the interior of North America, Greenland, and Northern Europe. The ancestral North American continent called Laurentia assembled from several microcontinents that collided beginning about 1.8 billion years ago. Most of the continent evolved in a relatively brief period of only 150 million years.

A major part of the continental crust underlying the United States from Arizona to the Great Lakes to Alabama formed in one great surge of crustal generation that has no known equal. This was possibly the most energetic period of tectonic activity and crustal generation in Earth history, when more than 80 percent of all continental mass was created. The best exposure of these Precambrian metamorphic rocks is the Vishnu Schist on the floor of the Grand Canyon (Fig. 85).

Laurentia was stable enough to resist another billion years of jostling and rifting. It continued to grow by plastering bits and pieces of continents and island arcs to its edges. The presence of large amounts of volcanic rock near the eastern edge of Laurentia implies the continent was once the core of a larger supercontinent. The central portion of the supercontinent was far removed from the cooling effects of subducting plates, where Earth's crust sinks into the mantle. As a result, the interior of the supercontinent heated and erupted with volcanism.

Figure 85 *Precambrian Vishnu Schist, Grand Canyon National Park, Arizona.*

(Photo by R. M. Turner, courtesy USGS)

TABLE 6 CLASSIFICATION OF VOLCANIC ROCKS

Property	Basalt	Andesite	Rhyolite
Silica content	Lowest about 50%, a basic rock	Intermediate about 60%	Highest more than 65%, an acid rock
Dark mineral content	Highest	Intermediate	Lowest
Typical minerals	Feldspar Pyroxene Olivine Oxides	Feldspar Amphibole Pyroxene Mica	Feldspar Quartz Mica Amphibole
Density	Highest	Intermediate	Lowest
Melting point	Highest	Intermediate	Lowest
Molten rock viscosity at the surface	Lowest	Intermediate	Highest
Formation of lavas	Highest	Intermediate	Lowest
Formation of pyroclastics	Lowest	Intermediate	Highest

After the rapid continent building, the interior of Laurentia experienced extensive igneous activity from 1.6 to 1.3 billion years ago. A broad belt of red granites and rhyolites, which are igneous rocks formed by molten magma solidifying below ground as well as on the surface (Table 6), extended several thousand miles across the interior of the continent from southern California to Labrador. The Laurentian granites and rhyolites are unique due to their sheer volume, suggesting the continent stretched and thinned almost to the breaking point. These rocks are presently exposed in Missouri, Oklahoma, and a few other localities. However, they are buried under sediments up to a mile thick in the center of the continent. In addition, vast quantities of molten basalt poured from a huge tear in the crust running from southeast Nebraska into the Lake Superior region about 1.1 billion years ago. Arcs of volcanic rock also weave through central and eastern Canada down into the Dakotas.

These massive outpourings of igneous rocks in the interior of the continent suggest that Laurentia was part of a supercontinent that formed about 1.6 billion years ago and broke up around 1.3 billion years ago, coinciding with the igneous activity. The supercontinent acted like an insulating blanket over the upper mantle, allowing heat to collect underneath it. About 1.1 billion years ago, vast quantities of molten basalt poured from a huge tear in the crust running from southeast Nebraska into the Lake Superior region.

Figure 86 *The Grenville orogeny in North America.*

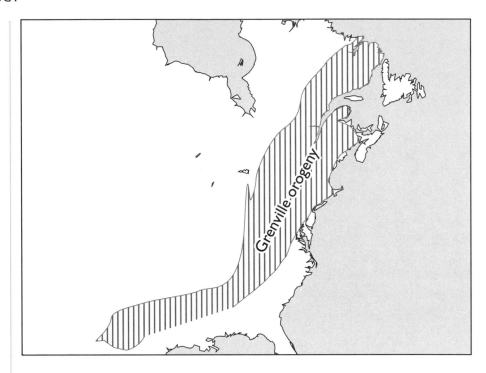

Figure 86 The Grenville orogeny in North America.

About 700 million years ago, Laurentia collided with another large continent on its southern and eastern borders, creating a new supercontinent called Rodinia. A superocean located approximately in the location of the present-day Pacific Ocean surrounded the supercontinent. The collision thrust up a 3,000-mile-long mountain belt in eastern North America called the Grenville orogeny (Fig. 86). A similar mountain belt occupied parts of western Europe as well. Around 670 million years ago, thick ice sheets spread over much of the landmass during perhaps the greatest period of glaciation Earth has ever known. At this time, Rodinia might have passed over one of the poles and collected a thick sheet of ice.

Rodinia rifted apart between about 630 and 560 million years ago, and its constituent continental blocks drifted away from one another. As the continents dispersed and subsided, seawater flooded the interiors, creating large continental shelves where vast arrays of organisms evolved. The rapid evolution of species at this time was highly remarkable. Another exceptional episode of explosive evolution occurred when a supercontinent called Pangaea rifted apart some 400 million years later.

The ancient North American continent was welded together from scraps of crust called cratons about 2 billion years ago. The African and South American continents did not aggregate until about 700 million years ago.

Over the past half-billion years, about a dozen individual continental plates came together to form Eurasia. It is the youngest and largest modern continent and is still being pieced together with chunks of crust arriving from the south, riding on highly mobile tectonic plates.

Rodinia rifted apart and the separated continents opened a proto–Atlantic Ocean called the Iapetus. The rifting process formed extensive inland seas, which submerged most of Laurentia some 540 million years ago, as evidenced by the presence of Cambrian seashores in such places as Wisconsin. It also flooded the ancient European continent called Baltica. The Iapetus was similar in location and size as the North Atlantic and was dotted with volcanic islands, resembling the present-day southwestern Pacific Ocean.

When the continents reached their maximum dispersal roughly 480 million years ago, subduction of the ocean floor beneath the North American plate initiated a period of volcanic activity and mountain building. From the early Silurian into the Devonian, about 420 million to 380 million years ago, Laurentia collided with Baltica, the ancient European landmass, and closed off the Iapetus. The collision fused the two continents into the supercontinent Laurasia (Fig. 87), named for the Laurentian province of Canada and the Eurasian continent. These Paleozoic continental collisions raised huge masses of rocks into several mountain belts throughout the world. The sutures joining the landmasses are preserved as eroded cores of ancient mountains.

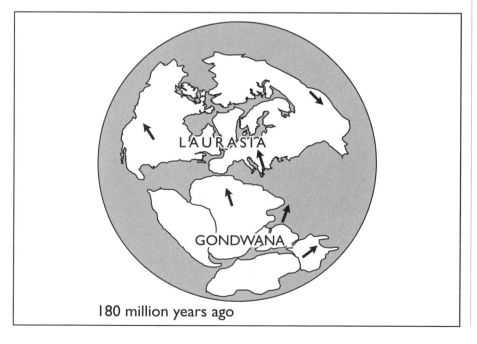

180 million years ago

Figure 87 *Distribution of the continents 400 million years ago with the formative Laurasia in the Northern Hemisphere and Gondwana in the Southern Hemisphere.*

When Laurentia united with Baltica, island arcs in a proto–Pacific Ocean called the Panthalassa began to collide with the western margin of what is now North America. The collisions led to the Antler orogeny, which intensely deformed rocks in the Great Basin region from the California-Nevada border to Idaho (Fig. 88). It is named for Antler Peak near Battle Mountain, Nevada. It consists of widespread, coarse clastic sediments and a prominent eastward-directed thrust fault known as the Roberts Mountain Thrust.

Laurasia occupied the Northern Hemisphere, while its counterpart Gondwana inhabited the Southern Hemisphere. Much of Gondwana was in the southern polar region from the Cambrian to the Silurian. The present continent of Australia sat on the equator at the northern edge of Gondwana. A large body of water called the Tethys Sea, named for the mother of the seas in Greek mythology, separated the two supercontinents. Evidence for the existence of a wide seaway between the landmasses comes from a unique specimen of flora called *Glossopteris* (Fig. 89) found in the southern lands but absent in Laurasia. The Tethys held thick deposits of sediments washed off the continents.

Figure 88 *The Crag Lake graben formed by block faulting in Idaho.*

(Photo by H. T. Stearns, courtesy USGS)

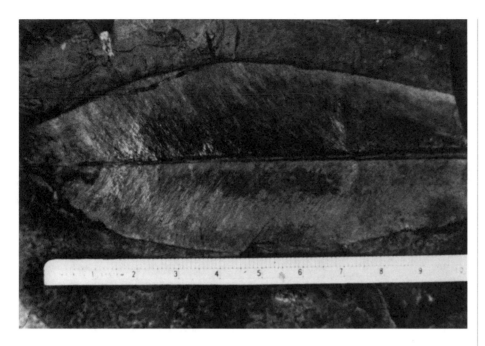

Figure 89 *Fossil*
Glossopteris *leaf, whose
existence on the southern
continents is strong
evidence for Gondwana.*

(Photo by D. L. Schmidt,
courtesy USGS)

The continents were lowered by erosion. Shallow seas flooded inland, covering more than half the present land area. The weight of the sediments formed a deep depression in the ocean crust called a geosyncline. The sedimentary rocks were later uplifted into great mountain belts surrounding the Mediterranean Sea when Africa slammed into Eurasia. The inland seas and wide continental margins along with a stable environment provided excellent growing conditions and enabled marine life to flourish and spread throughout the world.

During the late Silurian, Gondwana wandered into the southern polar region around 400 million years ago and acquired a thick sheet of ice. Glacial centers expanded in all directions. Ice sheets covered large portions of east central South America, South Africa, India, Australia, and Antarctica (Fig. 90). During the early part of the glaciation, the maximum glacial effects occurred in South America and South Africa. Later, the chief glacial centers switched to Australia and Antarctica, providing strong evidence that the southern continents wandered locked together over the South Pole.

In Australia, Silurian-age marine sediments were found interbedded with glacial deposits. Tillites, composed of glacially deposited boulders and clay, were separated by seams of coal. These indicate that periods of glaciation were punctuated by warm interglacial spells when extensive forests grew. The Karroo Series in South Africa is composed of a sequence of late Paleozoic lava

Figure 90 *Extent of late Paleozoic glaciation in Gondwana.*

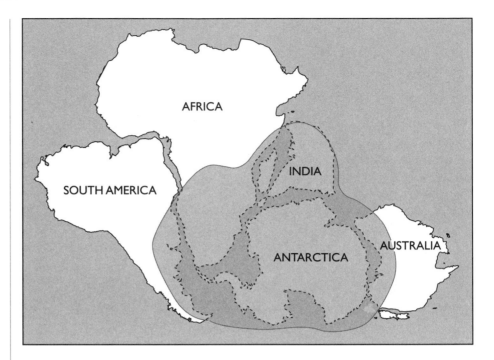

flows, tillites, and coal beds, reaching a total thickness of 20,000 feet. Between layers of coal were fossil leaves of the extinct fern *Glossopteris.* Because this plant is found only on the southern continents, it is among the best evidence for the existence of Gondwana.

After an examination of the earliest land plants of the Silurian, the next chapter will focus on the marine life of the Devonian period.

7

DEVONIAN FISH
THE AGE OF MARINE ANIMALS

This chapter examines life in the sea and the first vertebrates to come onto the land during the Devonian period. The Devonian, from 400 to 345 million years ago, was named for the marine rocks of Devon in southwest England. Rocks of Devonian age exist on all continents and show widespread marine and terrestrial conditions. The supercontinents Laurasia and Gondwana began approaching each other, pinching off the Tethys Sea between them. The wide distribution of deserts, evaporite deposits, coral reefs, and coal deposits as far north as the Canadian Arctic indicate a warm climate over large parts of the world.

The warm Devonian seas spurred the evolutionary development of marine species (Fig. 91), including the first appearance of the ammonoids. These were coiled-shelled cephalopods that became fantastically successful in the succeeding Mesozoic era. For 350 million years, species of these giant mollusks roamed the ancient seas. The vertebrates, the highest form of marine animal life, dominating all other creatures, left their homes in the sea to establish a permanent residence on land, which by then was fully forested. Toward the end of the Devonian, the climate cooled, possibly causing glaciation near the poles. The climate change caused the extinction of

Figure 91 *Marine fauna and flora of the middle Devonian.*

(Courtesy Field Museum of Natural History)

many tropical marine animals, paving the way for entirely new species especially adapted to the cold.

THE AGE OF FISH

The Devonian has been popularly named the "age of fish." The widespread distribution of primitive fish fossils throughout the world suggests a long vertebrate record early in the Paleozoic. The fossil record reveals so many and varied kinds of fish that paleontologists have a difficult time classifying them all. Every major class of fish alive today had ancestors in the Devonian. However, not all Devonian fish species continued to the present, having gone extinct in the intervening time.

The rise of fish in the Devonian seas contributed to the decline of their less mobile invertebrate competitors. This culminated in an extinction that eliminated many tropical marine groups at the end of the period. When a mass extinction occurs, those individuals that evolve into a better adaptive form are selected for survival, which is why certain species survive one major extinction after another. This is particularly true for marine species such as sharks, which originated in the Devonian around 400 million years ago and have survived every mass extinction since.

Fish comprise more than half the species of vertebrates, both living and extinct. They include the jawless fish (lampreys and hagfish), the cartilaginous fish (sharks, skates, rays, and ratfish), and the bony fish (salmon, swordfish, pickerel, and bass). The ray-finned fish are by far the largest group of living fish species. Fish progressed from rough scales, asymmetrical tails, and cartilage in their skeletons to flexible scales, powerful advanced fins and tails, and all-bone skeletons, much like they are today.

The jawless fish, which first appeared in the Ordovician, are the earliest known vertebrates, having been in existence for 470 million years. They had a flexible rod similar to cartilage, which functioned as a spine along the back. They were probably poor swimmers, however, and restricted to shallow water. Bony plates surrounded the head for protection from invertebrate predators. However, the additional weight forced the fish to live mostly on the seafloor, where they sifted bottom sediments for food particles.

The development of jaws about 460 million years ago revolutionized predation. Giant jawed vertebrates, some of which were monsters in their day, climbed to the very top of the food chain. The extinct placoderms (Fig. 92), which reached 30 or more feet in length, were ferocious giants that preyed on smaller fish. They lived in shallow, freshwater streams and lakes. They had camouflage of red scales that helped them blend in well with their reddish brown habitat. They had well-developed articulated jaws and thick armor plating around the head that extended over and behind the jaws. One of these groups gave rise to land animals, emphasizing the great importance jaws played in vertebrate evolution.

Figure 92 *The extinct placoderms were giants measuring 30 feet in length.*

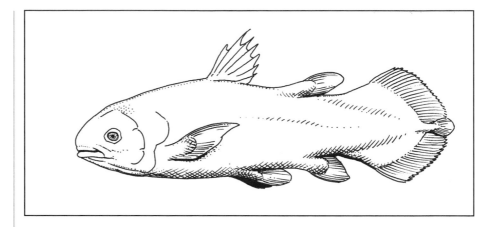

The evolution of jaws also improved fish respiration by supporting the gills. After a fish draws water into its mouth, it squeezes the gill arches to force the water over the gills at the back of the mouth. Blood vessels in the gills exchange oxygen and carbon dioxide as the water flows out the gill slits. The jaws had the advantage of clamping down on significantly large prey, allowing fish to become fierce predators. Primitive jawed fish might have even caused the demise of the trilobites, once spectacularly successful in the Cambrian seas.

The coelacanths (Fig. 93) were thought to have gone extinct along with the dinosaurs 65 million years ago. However, in 1938, fishermen caught a 5-foot coelacanth in the deep, cold waters of the Indian Ocean off the Comoro Islands near Madagascar. The fish looked ancient, a castaway from the distant past. It had a fleshy tail, a large set of forward fins behind the gills, powerful square toothy jaws, and heavily armored scales. The most remarkable aspect about this fish was it had not changed significantly from its primitive ancestors, which evolved in the Devonian seas some 400 million years earlier. Because of this, the coelacanth has been given the title of "living fossil."

The coelacanth's head contained a small organ that is thought to detect faint electric fields. Sharks have similar sensors to home in on weak electric fields generated by the moving muscles of smaller fish upon which they prey. The coelacanth would perform a number of acrobatic feats, including headstands, swimming backward, or flipping upside down to pinpoint the electric tracks of prey.

The coelacanth came from the same evolutionary branch in direct line to land-dwelling vertebrates. Stout fins on the fish's underside enabled it to crawl along the deep ocean floor. The fins were precursors of amphibian limbs and were coordinated in a manner not seen in most fish but common in four-legged terrestrial animals. The fins moved similarly to the legs of a crawling lizard, with the forward appendage on each side advancing in concert with the rear appendage on the opposite side. Such an adaptation would have eased the transition from sea to land, making the coelacanth the most direct ancestor of higher terrestrial animals.

A possible link between fish and terrestrial vertebrates were the Devonian crossopterygians and lungfish, another living fossil still in existence today. The crossopterygians were lobe finned, meaning that the bones in their fins were attached to the skeleton and arranged into primitive elements of a walking limb. They breathed by taking air into primitive nostrils and lungs as well as by using gills. This placed them into the direct line of evolution from fish to land-living vertebrates that gave rise to amphibians and reptiles (Fig. 94).

The sharks were highly successful from the Devonian to the present. An ancient freshwater shark called *Xenacanthus* had a back fin that stretched from head to tail, allowing it to slither through the water like an aquatic snake. Closely related to the sharks are the rays, with flattened bodies, pectoral fins enlarged into wings up to 20 feet across, and a tail reduced to a thin, whiplike appendage. The rays literally fly through the sea as they scoop up plankton into their mouths.

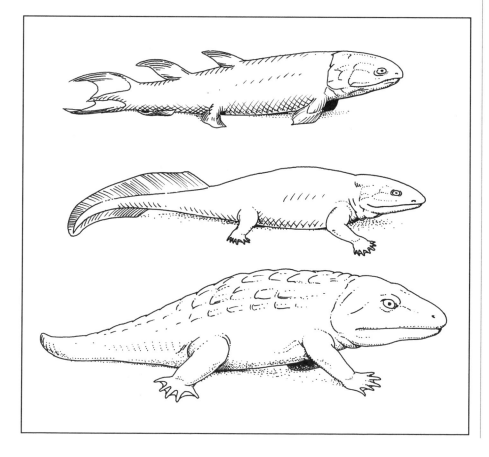

Figure 94 The rough evolution from crossopterygians (top) *to the amphibious fish* (middle) *to the amphibians* (bottom).

Figure 95 *Excavation for shark teeth at Shark's Tooth Hill, Kern County, California.*

(Photo by R. W. Pack, courtesy USGS)

Sharks breathe by drawing water in through the mouth, passing it over the gills, and expelling it through distinctive slits behind the head. The body of the shark is heavier than water, requiring it to swim constantly or else sink to the bottom. Instead of skeletons composed of bone as with most fish, shark skeletons are made of cartilage, a much more elastic and lighter material. However, cartilage does not fossilize well. About the only common remains of ancient sharks are teeth, found in marine rocks of Devonian age onward (Fig. 95).

MARINE INVERTEBRATES

The Devonian marine invertebrates were similar to those that evolved in the Ordovician and include prolific brachiopods, corals, crinoids, trilobites, and gastropods. The more advanced articulate brachiopods appeared in the Devonian and became important stratigraphic markers for the period. Rocks from the Cambrian and Devonian periods contain brachiopod fossils and wave marks, indicating that some ancient forms inhabited the shore areas. Modern forms, which number about 260 species, inhabit warm ocean bottoms from a few feet to more than 500 feet deep. Some rare types thrive at depths approaching 20,000 feet.

Brachiopod shells are lined on the inside of the valves with a membrane called a mantle. This encloses a large central cavity that holds the lophophore, which functions in food gathering. Projecting from a hole in the valve is a muscular stalk called a pedicel by which the animal is attached to the seabed. The structure of the valves aids in the identification of various brachiopod species. The shells come in a variety of forms, including ovoid, globular, hemispherical, flattened, convex-concave, or irregular. The surface is smooth or ornamented with ribs, grooves, or spines. Growth lines and other structures show changes in form and habit that offer clues to brachiopod history.

The bryozoans were tiny coral-like colonies, with an encrusting, branching, or fanlike structure. They were a major group of marine fauna that witnessed a marked change from early to late Paleozoic. The stony bryozoans, which were particularly abundant in the Ordovician and Silurian, declined to insignificance by the Devonian. The lacy forms greatly diversified and dominated all bryozoan groups, reaching a peak in the Devonian and Carboniferous. Delicate, twiglike bryozoans were also common but declined significantly in the Permian, with only a few groups surviving the extinction at the end of the period.

The tabulate corals that dominated the Ordovician and Silurian were not nearly as abundant in the Devonian. The tetracorals, which had fourfold asymmetry with septa arranged in quadrants, were an important group of reef-building corals that reached their peak during the Devonian. They built the reef that forms the Falls of the Ohio River at Louisville, Kentucky. The tetracorals were still common in the Carboniferous but declined significantly in the Permian, as the seas in which they thrived receded. They possibly gave rise to the hexacorals, with six-sided sepia. Most tabulate corals and tetracorals failed to survive the Permian extinction.

The glass sponges, with an interlocking gridwork of siliceous spicules, were very common in the shallow Devonian seas. They consisted of glasslike fibers of silica intricately arranged to form a beautiful network. These hard skeletal structures are generally the only parts of sponges preserved as fossils. The great success of the sponges along with organisms such as diatoms that extract silica directly from seawater to construct their skeletons explains why today's ocean is largely depleted of this mineral.

The conodonts (Fig. 96), which are bony appendages of a possible leech-like animal that had the appearance of jawbones, show their greatest diversity

Figure 96 *The conodonts show their greatest diversity in the Devonian.*

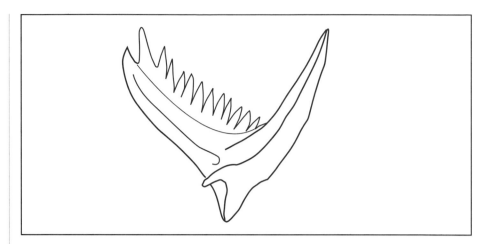

Figure 96 *The conodonts show their greatest diversity in the Devonian.*

during the Devonian. They are among the most common fauna and important for long-range rock correlations of this period. The conodonts were of lesser importance in the Permian and Triassic, when the shallow seas that they inhabited became highly restrictive.

The mollusks were well represented. The first appearance of freshwater clams suggests that aquatic invertebrates had successfully conquered the land by the Devonian. Arthropods also invaded the land. They included advanced insects, spiders, and centipedes with poisonous jaws, which first appeared in the Devonian. Ostracods, tiny bivalved crustaceans also known as mussel shrimp, increased in abundance but declined in the late Permian when the seas contracted due to continental collisions. The flowerlike crinoids were common in the Devonian seas, reaching a peak in the late Paleozoic. However, the crinoids and their blastoid cousins failed to survive past the Permian.

The nautiloids and ammonoids (Fig. 97) appeared in the early Devonian about 395 million years ago. They had external shells subdivided into air chambers. The suture lines joining the segments presented a variety of patterns used for identifying various species. The air chambers provided buoyancy to counterbalance the weight of the growing shell. Most shells were coiled in a plane, some forms were spirally coiled, and others were essentially straight.

The now-extinct nautiloids grew upward of 30 or more feet long. With their straight, streamlined shells, they were among the swiftest and most spectacular creatures of the Devonian seas. Their neutral buoyancy, maintained by air chambers, and jet propulsion for high-speed travel, by expelling water under pressure through a funnel-like appendage, contributed to the nautiloid's great success.

The belemnoids, which probably originated from more primitive nautiloids, were abundant during the Jurassic and Cretaceous but became extinct by the Tertiary. They were related to the modern squid and octopus and possessed a long, bulletlike shell. The shell was straight in most species and loosely coiled in others. The chambered part of the shell was smaller than the ammonoid, and the outer walls thickened into a fat cigar shape.

The ammonoids were the most significant cephalopods. They had a large variety of coiled shell forms (Fig. 98), which made them ideal for dating Paleozoic and Mesozoic rocks. Shell designs steadily improved, making ammonoids the swiftest creatures of the deep. They successfully competed with fish for food and avoided predators. Ammonoids lived mainly at middle depths and might have shared many features with living squids and cuttlefish. Some ammonoids grew to tremendous size with shells up to 7 feet wide. The nautilus, which is commonly referred to as a living fossil because it is the only living relative of the ammonoids, lives in the depths of the South Pacific and Indian oceans down to 2,000 feet.

During a major extinction event near the end of the Devonian about 365 million years ago, many tropical marine groups disappeared, possibly due to climatic cooling. Several extinctions correlate with glaciations. Yet no major die out occurred during the widespread Carboniferous glaciation, which enveloped the southern continents around 330 million years ago. The relatively low extinction rates were credited to the limited number of extinction-prone species following the late Devonian extinction.

The die out of species at the end of the Devonian apparently occurred over a period of 7 million years. It eliminated species of corals and many other bottom-dwelling marine organisms. Primitive corals and sponges, which were prolific limestone reef builders early in the period, suffered heavily during the extinction and never fully recovered. While these groups vanished, the glass sponges, which better tolerated cold conditions, rapidly diversified, only to

Figure 97 The ammonoids were among the most spectacular creatures of the Paleozoic and Mesozoic seas, with some growing up to 7 feet across.

Figure 98 *A collection of ammonoid shells from a formation in Arkansas.*

(Photo by M. Gordon Jr., courtesy USGS)

dwindle when the crisis subsided and other groups recovered. Their prosperity during the late Devonian signifies that less fortunate species had succumbed to the effects of climatic cooling. Large numbers of brachiopod families also died out at the end of the period.

The extinction did not identically affect all species that shared the same environments, however. Many Gondwanan fauna survived the onslaught due to the scarcity of reef builders and other warm-water species prone to extinction.

In contrast, cold–adapted animals living in Arctic waters fared quite well. Much of Gondwana was in the Antarctic during the Devonian, and seas flooded broad areas of the continent. The Gondwanan fauna, which lacked reef builders and other warm–water species, survived the extinction with few losses.

The oldest species living in the world's oceans today thrive in cold waters. Many Arctic species, including certain brachiopods, starfish, and bivalves, belong to biological orders whose origins extend hundreds of millions of years back to the Paleozoic. In contrast, tropical faunas such as reef communities, battered by periodic mass extinctions, have come and gone quite rapidly on the geologic time scale. However, not all animals that shared the same environments such as corals and mollusks were identically affected by the extinctions.

A possible cause for the end–Devonian extinction was the bombardment of Earth by one or two large asteroids or comets. The meteorite impact theory is supported by the discovery of deposits containing glassy beads called tektites in the Hunan province of China and in Belgium. Tektites form when a large meteorite strikes Earth and hurls droplets of molten rock into the air that quickly cool into bits of glass (Fig. 99). The deposits also include an unusually high iridium content, which strongly indicates an extraterrestrial source. The Siljan crater in Sweden, about the same age as the tektites, might be the source

Figure 99 *A North American tektite found in Texas in November 1985, showing surface erosional and corrosional features.*

(Photo by E. C. T. Chao, courtesy USGS)

Figure 100 *Ancient ferns such as* Medullosa *were as tall as present-day trees.*

of the impact deposits. The evidence suggests that meteorite bombardments might have contributed to many mass extinctions throughout Earth history.

TERRESTRIAL VERTEBRATES

The Devonian floral landscape was dominated by the true ferns, which are the second most diverse group of living plants. The whisk ferns, which appeared at the end of the Silurian, became extinct near the end of the Devonian. Some ancient ferns attained heights of present-day trees (Fig. 100). The earliest woody tree, the extinct *Archaeopteris,* dramatically transformed the planet in the late Devonian more than 370 million years ago, when amphibians were just beginning to craw out of the sea onto the land. The tree produced spores and had thick permanent deciduous branches as well as short-lived branches that dropped off after a couple of years.

The thick branches would have shaded and cooled the streams where amphibians were rapidly evolving. As plants spread across the globe, the burgeoning vegetation drew carbon dioxide out of the atmosphere thereby cooling the planet while simultaneously adding oxygen to the air and setting the stage for land vertebrates. At the beginning of the Devonian, land animals would have been unable to breathe due to the low atmospheric oxygen content. However, by the end of the period, they would not have had any difficulty breathing.

Freshwater invertebrates and fish inhabited lakes and streams. Freshwater fish living in Australia around 370 million years ago were almost identical to those living in China. This suggests the two landmasses were close enough for the fish to travel between them. Fish had remained the only vertebrates until around 360 million years ago, with the evolution of the lobe-finned fish such as the crossopterygians (Fig. 101). The lobe-finned fish had thick, rounded fins whose bones were crude forerunners of those in tetrapod (four-legged) limbs. Lobe-finned fish used gills for respiration. They could also breathe with primitive lungs in oxygen-poor swamps or when stranded on dry land. Their descendants became the first advanced animals to populate the continents.

The ancestors of the amphibians appear to have been the crossopterygians, the stem group from which all terrestrial vertebrates descended. They grew upward of 10 feet long and had powerful jaws containing large teeth. The amphibians began to dominate the land by the middle Devonian and were especially attracted to the great swamps. When the climate chilled and glaciers spread over the continents at the end of the period, the first reptiles emerged and began to displace the amphibians as the dominant land vertebrates.

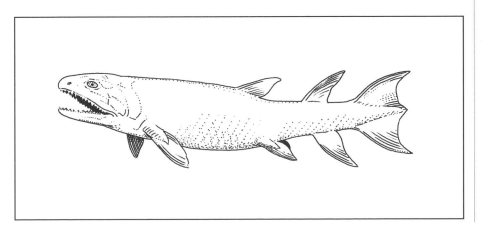

Figure 101 Lobed-finned fish evolved into the first tetrapods.

Figure 102 *Air-breathing fish traveled overland to new water holes.*

Lobe-finned fish breathed with gills. Unlike other fish, they also breathed with lungs. These were the predecessors of modern lungfish. An abundance of food swept up onto the beaches during high tide might have enticed these fish to come ashore. Fierce competition in the ocean for scarce food supplies provided an extraordinary evolutionary incentive for any animal that could find food on land. Their descendants became the first advanced animals to populate the continents.

By the middle Devonian, stiff competition in the sea encouraged crossopterygians to make short forays onto shore to prey on abundant crustaceans and insects. The crossopterygians were lobe-finned fish with heavy, enamel-like scales. Their fin bones were attached to the skeleton in such a manner to form primitive limbs. The crossopterygians strengthened their lobe fins, which eventually evolved into legs, by digging in the sand for food and shelter. They eventually ventured farther inland, though not too distant from accessible sources of water such as swamps or streams. Primitive Devonian fish, similar to today's lungfish, crawled on their bellies from one pool to another, pushing themselves along with their fins (Fig. 102).

Links between fish and terrestrial vertebrates were the Devonian lungfish. Their descendants still live today in Africa, Australia, and South America, which comprised the supercontinent Gondwana. Modern lungfish live in African swamps that seasonally dry out, forcing the fish to hole up for long stretches until the rains return. They burrow into the moist sand, leaving an air hole to the surface, and live in suspended animation, breathing with primitive lungs. In this manner, they can survive out of water for several months or even a year or more if necessary. When the rainy season returns, the pond fills again, and the fish come back to life, breathing normally with their gills.

In Florida, a walking catfish originating from Asia will leave its drying pond and travel by pushing itself along with its tail and fins, sometimes a considerable distance before finding another suitable home. It breathes with primitive nostrils and lungs as well as with gills, placing it midway on the line of evolution from fish to land-living vertebrates. Air breathing is also important for fish striving to survive in warm, shallow, stagnant waters with a low oxygen content.

The descendants of the lobe-finned fish and lungfish were the first advanced animals to populate the land some 370 million years ago. By the late Devonian, the descendants of the crossopterygians evolved into the earliest amphibians. Their legacy is well documented in the fossil record. At no other time in geologic history were so many varied and unusual creatures inhabiting the surface of Earth.

Animal tracks tell of the earliest land invasion. Tracks of primitive Devonian fish that first ventured onto dry land and gave rise to the four-legged amphibians exist in formations of late Devonian age onward. Amphibian footprints became abundant in the Carboniferous beginning about 350 million years ago and to a lesser extent in the Permian, owing to the amphibians' preference for a life in water and to the rise of the reptiles. The fossil remains of the amphibians are largely fragmentary, however, because of the manner by which vertebrate skeletons are constructed. Their large number of bones that are easily scattered by surface erosion leave a scant record of their existence.

THE OLD RED SANDSTONE

Beginning in the late Silurian and continuing into the Devonian, from about 400 million to 350 million years ago, a collision between present eastern North America and northwestern Europe raised the Acadian Mountains (Fig. 103). The terrestrial red beds of the Catskills in the Appalachian Mountains of southwestern New York State to Virginia are composed of sandstones and shales cemented by red iron oxide and are the main expression of the Acadian orogeny in North America. Extensive igneous activity and metamorphism accompanied the mountain building at its climax.

The Devonian Antler orogeny was another mountain-building episode, resulting from a collision of island arcs with the western margin of North America. The island arcs appear to have formed approximately 470 million years ago off the west coast of North America. The orogeny intensely deformed rocks in the Great Basin region from the California-Nevada border to Idaho.

The Innuitian orogeny, from the Devonian through the Carboniferous, deformed the northern margin of the present North American continent. The mountain building episode raised the Innuitian Mountains on Ellesmere Island in the Canadian Arctic. It resulted from a collision with another crustal plate, possibly the eastern Siberian continental mass. Block faulting and basin filling succeeded the mountain building in the region.

Figure 103 *Location of the ancient Acadian Mountains in North America.*

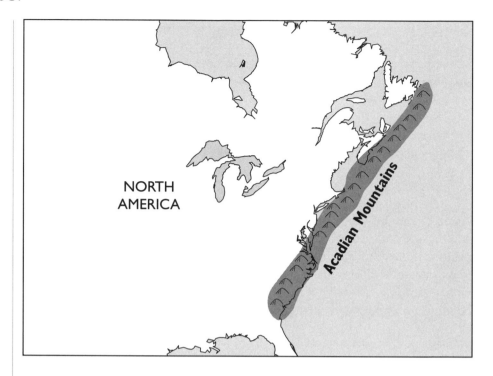

The middle Devonian Old Red Sandstone, a thick sequence of chiefly nonmarine sediments in Great Britain and northwestern Europe (Fig. 104), is the main expression of a mountain building episode called the Caledonian orogeny. The formation comprises great masses of sand and mud that accumulated in the basins between the ranges of the Caledonian Mountains from Great Britain to Scandinavia. The sediments are poorly sorted (dissimilar in size) and consist of red, green, and gray sandstones and gray shales that often contain fish fossils.

Erosion leveled the continents and shallow seas flowed inland, flooding more than half the landmass. The inland seas and wide continental margins along with a stable environment provided favorable conditions for marine life to flourish and proliferate throughout the world. Seas flooding North America during the Devonian produced abundant coral reefs that lithified (became rock) into widespread limestones (Fig. 105).

The rising Acadian Mountains on the east side of the inland sea eroded away. Their sediments produced flat-lying, fossiliferous deposits of shale in western New York State, possibly the best Devonian section in the world. The vast Chattanooga Shale Formation, which covers virtually the entire continental interior, was laid down during the Devonian and Carboniferous. The seas also blanketed much of Eurasia in the late Devonian. Terrestrial clastics containing rock fragments eroded from the Caledonian Mountains overlay the western part of the continent.

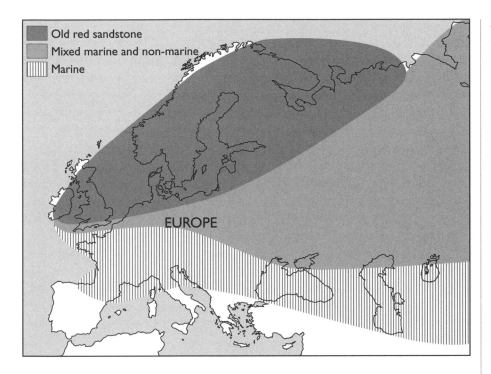

Figure 104 *Location of the Old Red Sandstone in Northern Europe.*

Figure 104 *Location of the Old Red Sandstone in Northern Europe.*

Figure 105 *A limestone formation of the Bend Group of the Sierra Diablo Escarpment, Culberson County, Texas.*

(Photo by P. B. King, courtesy of USGS)

The second half of the Paleozoic followed a Silurian ice age, when Gondwana wandered into the southern polar region around 400 million years ago and acquired a thick sheet of ice. Gondwana, to this point located in the Antarctic, now shifted its position. Its location can be shown by paleomagnetic data, which indicate the locations of continents relative to the magnetic poles by analyzing the magnetic orientations of ancient iron-rich lavas. The south magnetic pole drifted from present South Africa in the Devonian, ran across Antarctica in the Carboniferous, and ended up in southern Australia in the Permian.

The location of the southern pole is also indicated by widespread glacial deposits and erosional features on the continents that comprised Gondwana during the late Paleozoic. The mass extinctions of the late Ordovician 440 million years ago and the middle Devonian 365 million years ago coincided with glacial periods that followed long intervals of ice-free conditions.

Gondwana in the Southern Hemisphere and Laurasia in the Northern Hemisphere were separated by the Tethys Sea (Fig. 106). Into this seaway flowed thick deposits of sediments washed off the surrounding continents. Their accumulated weight formed a long, deep depression in the ocean crust, called a geosyncline, which later uplifted into folded mountain belts when Gondwana and Laurasia collided.

A warm climate and desert conditions over large areas are indicated by the widespread distribution of evaporite deposits in the Northern Hemisphere, coal deposits in the Canadian Arctic, and carbonate reefs. Warm temperatures of the past are generally recognized by abundant marine limestones, dolomite, and calcareous shales. A coal belt, extending from northeastern Alaska across the Canadian archipelago to northernmost Russia, suggests that vast swamps were prevalent in these regions.

Figure 106 *Around 400 million years ago, all continents surrounded an ancient sea called the Tethys.*

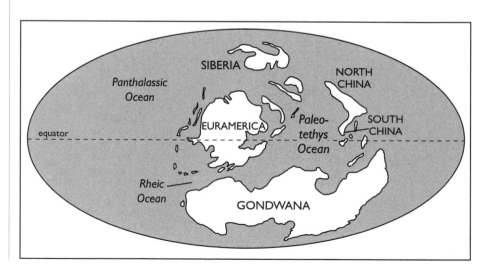

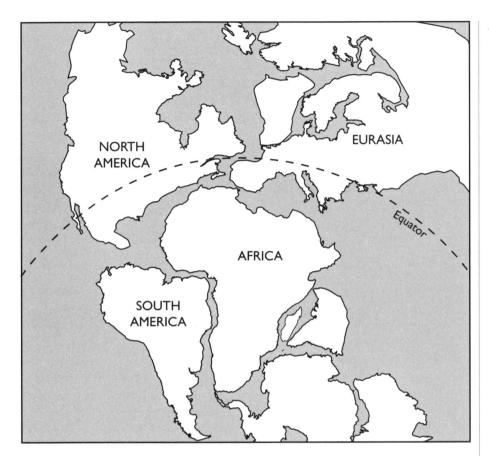

Figure 107 The approximate positions of the continents relative to the equator during the Devonian and Carboniferous periods.

Evaporite deposits generally form under arid conditions between 30 degrees north and south of the equator. However, extensive evaporite deposits are not currently being formed, suggesting a comparatively cooler global climate. The existence of ancient evaporite deposits as far north as the Arctic regions implies that either these areas were once closer to the equator or the global climate was considerably warmer in the geologic past.

The Devonian year was 400 days long and the lunar cycle was about 30.5 days as determined by the daily growth rings of fossil corals. Paleomagnetic studies indicate that the equator passed from California to Labrador and from Scotland to the Black Sea during the Devonian and Carboniferous (Fig. 107). The ideal climate setting helped spur the rise of the amphibians that inhabited the great Carboniferous swamps.

After covering the marine and terrestrial life-forms of the Devonian, the next chapter follows the evolution of the first amphibians of the Carboniferous period.

8

CARBONIFEROUS AMPHIBIANS
THE AGE OF FOREST DENIZENS

This chapter examines the evolution of the amphibians in the great coal swamps of the Carboniferous period and follows the building of the great supercontinent Pangaea. The Carboniferous, from 345 to 280 million years ago, was named for the coal-bearing rocks of Wales, Great Britain. It is further divided into the Mississippian and Pennsylvanian periods in North America. Flora that appeared in the Devonian was plentiful and varied during the Carboniferous. Great coal forests of seed ferns and true trees with seeds and woody trunks spread across Gondwana and Laurasia in the lower Carboniferous.

All forms of marine fauna that existed in the lower Paleozoic flourished in the Carboniferous except the brachiopods, which declined in number and types. The fusulinids appeared for the first time in the Carboniferous. They were large, complex protozoans that resembled grains of wheat and ranged from microscopic size up to 3 inches in length. Primitive amphibians inhabited the swampy forests, which were abuzz with hundreds of different types of insects, including large cockroaches and giant dragonflies. When the climate grew colder and widespread glaciation enveloped the southern continents at the end of the period, the first reptiles emerged and displaced the amphibians as the dominant land vertebrates.

THE AMPHIBIAN ERA

Plants had been greening Earth for as long as 100 million years before the vertebrates finally set foot onto dry land. Prior to the amphibian invasion, freshwater invertebrates and fish had been inhabiting lakes and streams. By the early Carboniferous, the vertebrates had spent more than 160 million years underwater, with only a few short forays onto the land. Relatives of the lungfish lived in freshwater pools that dried out during seasonal droughts, requiring the fish to breathe with primitive lungs as they crawled to the safety of the nearest water hole.

A common misconception is that land animals evolved from fish and developed armlike limbs after flopping onto shore. A distant ancestor of the catfish living 370 million years ago had complex fingerlike bones in the underside of its fin and spent its life entirely in water. The fish grew to 8 feet long and weighed up to 200 pounds. It dated from within a few million years of the beginning of the tetrapods, the first four-legged animals with vertebrae that were the forebears of the amphibians and reptiles.

The fish had complex eight-fingered, handlike limbs, suggesting that some fish developed legs first before venturing out onto land. The fingered fin shows that limb bones could have evolved in fish for use in water long before being useful on land. The fish probably used its walking fins to help it move around and hunt in the shallow, weedy streams, giving it a distinct advantage over other fish. Also, at this time, shallow swamps began to appear. Limbs with digits would have helped the first aquatic tetrapods travel through the plant-choked wetlands. Eventually, the fins evolved into paws that its progeny could use on land.

Amphibious fish (Fig. 108) probably spent little time on shore because their primitive legs could not support their body weight for long periods, requiring them to return to the water. Eventually, as their limbs strengthened, the amphibious fish wandered farther inland, where crustaceans and insects were abundant. By the middle Devonian, they began to dominate the land and were especially attracted to the great Carboniferous swamps.

By about 335 million years ago, the amphibious fish evolved into the earliest amphibians. The tetrapods branched into two groups, with one line leading to amphibians and the other to reptiles, dinosaurs, birds, and mammals. Some species had strong, toothy jaws and resembled giant salamanders, reaching 3 to 5 feet in length. A 2-foot-long amphibian with armadillo like plates rooted in the soil for worms and snails. During the early Carboniferous, the

Figure 108 *Amphibious fish were the first tetrapods.*

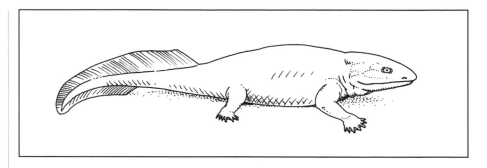

land offered no large animals to eat, and vertebrates had not yet evolved the capacity to consume plants. Therefore, the only food consisted of invertebrates, including millipedes, centipedes, and the forerunners of insects.

The earliest known tetrapod was *Acanthostega* (Fig. 109), meaning "spine plate." It was essentially an aquatic animal with attached hands and feet. The salamander-like body had large eyes on top of a flat head for spotting prey swimming above as it sat buried in the bottom mud. It sported eight toes on the front feet and seven toes on the rear feet, perhaps the most primitive of walking limbs. The digits were sophisticated and multijointed. However, because they were attached to an insubstantial wrist, the legs were virtually useless for walking on the ground. The rest of the skeletal anatomy also suggests acanthostega could not have easily walked on land. Instead, it probably crawled around on the bottom of lagoons and used gills for respiration.

One of the earliest known amphibians was an ancient land vertebrate called *Ichthyostega,* meaning "fish plate," which lived its life half the time in

Figure 109
Acanthostega *walked on the bottoms of lagoons with legs having eight toes.*

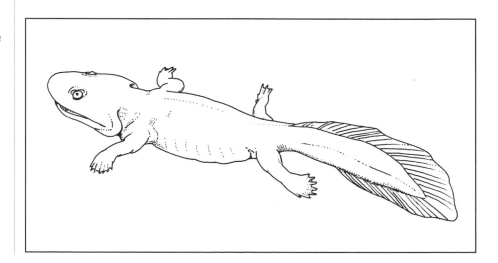

water and half the time on dry land. It was dog sized, with a broad, flat, fish-like head and a tail topped with a small fin, apparently used for swimming. It developed a sturdy rib cage to hold up its internal organs while on land and crawled around on primitive legs with seven toes on the hind limbs. Amphibians also possessed six and eight digits on their feet, indicating the evolution of early land vertebrates followed a flexible pattern of development. However, no terrestrial vertebrates evolved a foot with more than five true digits.

Neither acanthostega nor ichthyostega could do much more than waddle around on land. Their upper arm bones had a broad, blobby shape ill suited for walking. Their hind limbs splayed out to the side and could not have easily held up the body. The backbones were looser than those of terrestrial tetrapods and were similar to those of fish, which offered less support on land.

A small amphibianlike animal called microsaur was less than 6 inches long, a mere midget compared with its aquarian counterparts. It was among the first four-legged vertebrates to crawl onto the land more than 300 million years ago. The animal had uniquely shaped spine bones and a simplified skull, with just one bone instead of three. The skull was attached to the first vertebra of the spine with a pair of rodlike bones, which limited head rotation only to up-and-down movements. This feature is also found in the large, more primitive amphibians that predated microsaur.

The weak legs of the early amphibians could hardly keep their squat bodies off the ground, making them slow and ungainly. The amphibian tracks are generally broad with a short stride. The animal walked with a clumsy gait. Therefor, running to attack prey or escape predators was simply not possible. In order to succeed as hunters without requiring speed or agility, the amphibians developed a unique whiplike tongue that lashed out at insects and flicked them into the mouth. This successful adaption enabled the amphibians to populate the land rapidly.

Although the amphibians had well-developed legs for walking on dry land, the animals apparently spent most of their time in rivers and swamps. They depended on accessible sources of water to moisten their skins as well as for respiration and reproduction. They reproduced like fish, laying small, shell-less eggs. After hatching, the juveniles lived an aquatic, fishlike existence, breathing with gills. As they matured, the young amphibians metamorphosed into air-breathing, four-limbed adults.

The early amphibians living during the late Devonian, when vertebrates were first making a transition from sea to land, spent most of their time in the water. The necessity of having to live a semiaquatic lifestyle, however, led to the eventual downfall of the amphibians when the great swamps began to dry out toward the end of the Paleozoic. The void left by the amphibians was quickly filled by their cousins the reptiles, which were better suited for a life totally out of water.

Amphibian footprints were quite abundant during Carboniferous period but less so in the Permian, owing to the takeover by the reptiles and the amphibians' preference for life in the water. The increase in the number of reptilian footprints in the Carboniferous and Permian plainly shows the rise of the reptiles at the expense of the amphibians. Possibly one of the major factors leading to the superiority of the reptiles was their more efficient mode of locomotion. The reptiles were also much more suited for living full-time on dry land, whereas the amphibians had to return to the water periodically.

Populations of amphibians continued to fall during the Mesozoic, with all large, flat-headed species going extinct. The group thereafter was represented by the more familiar salamanders, toads, and frogs. The fossil remains of these amphibians are largely fragmentary because vertebrate skeletons are constructed with a large number of bones that are easily scattered by surface erosion. Although the amphibians did not achieve complete dominion over the land, their cousins the reptiles were destined to become the greatest success story the world has ever known.

THE GREAT COAL FORESTS

During the second half of the Paleozoic, the continents rose and sea levels dropped. This caused the departure of the inland seas, which were replaced with immense swamps. About 315 million years ago, extensive forests grew in the great swamps. These regions formed a vast tropical belt that ran through the supercontinent Pangaea, which straddled the equator.

By the middle Paleozoic, the terrestrial flora was plentiful and varied. The most significant evolutionary step was the development of a vascular stem to conduct water to a plant's extremities. The early club mosses, ferns, and horsetails were the first plants to utilize this water vascular system. The early complex land plants diverged into two major groups. One gave rise to the lycopods. The other spawned the gymnosperms, which were ancestral to many modern land plants. The gymnosperms, including cycads, ginkgos, and conifers, originated in the Permian and bore seeds that lacked fruit coverings.

Eventually, as evolution steadily progressed, great coal forests spread across the continents. The woodlands included thick stands of seed ferns and true trees, which were gymnosperms with seeds and woody trunks. By about 370 million years ago, forests transformed the planet with green, lush vegetation. The lycopods ruled the ancient swamps. They towered as high as 130 feet and became the first trees to develop true roots and leaves, which were generally small. Branches were arranged in a spiral. Spores were attached to modified leaves that became primitive cones. Their trunks were largely composed of bark and, for the most part, lacked branches along the length of the tree, looking much like a forest of telephone poles. Only near the end of their lives did the lycopods sprout a small crown of limbs as they prepared for reproduction.

Making their living among these trees were giant insects, huge millipedes, walking fish, and primitive amphibians. The ancestors of the dragonfly with 3-foot wingspans, swallow-sized mayflies, and other monstrous insects ruled the skies 300 million years ago. In one of the longest-running battles on Earth, plants and insects have fought each other for more than 300 million years. Some of the fiercest wars were staged in the tropical regions. Hungry insects fiercely attacked vegetation, which protected itself with multiple defensive weapons, including chemical warfare.

The great success of the insects was probably due to an abundance of atmospheric oxygen. During the Carboniferous, oxygen levels might have reached 35 percent of Earth's atmosphere as compared with today's 21 percent. Growing big was an insect's way of taking advantage of the oxygen-rich air. The higher oxygen levels might have given the early amphibians a chance to develop their lungs and become permanently established on land. However, by 245 million years ago, the oxygen heyday was over. Levels dropped to 15 percent, which might have contributed to the massive extinctions at the end of the Permian.

For millions of years, the lycopods endured changes in sea level and climate that alternately drained and flooded the swamps. Then about 310 million years ago, the climate of the tropics became drier and most swamplands disappeared entirely. The climate change set off a wave of extinctions that wiped out virtually all lycopods at the beginning of the Permian 280 million years ago. Today, they exist only as small grasslike plants in the tropics. Late in the Carboniferous, as the climate grew wetter and the swamps reemerged, weedy plants called tree ferns (Fig. 110) dominated the Paleozoic wetlands.

The second most diverse group of living plants were the true ferns. They ranged from the Devonian to the present. However, they were particularly widespread in the Mesozoic and prospered well in the mild climates even in the higher latitudes. In contrast, today they are restricted to the tropics. Some ancient ferns attained heights of present-day trees. The Permian seed fern *Glossopteris* was especially significant. Its fossil leaves are prevalent on the continents that formed Gondwana but are lacking on the continents that comprised Laurasia. This indicates that these two landmasses were in separate parts of the world divided by the Tethys Sea. This body of water was wide in the east and narrow in the west, where land bridges aided the migration of plants and animals from one continent to the other.

Terrestrial fossils are not nearly as abundant as those of marine origin, primarily because land species do not fossilize well and fossil-bearing sediments are subjected to erosion. However, some environments such as swamps and marshes provided an abundance of plant and animal fossils. Well-preserved,

147

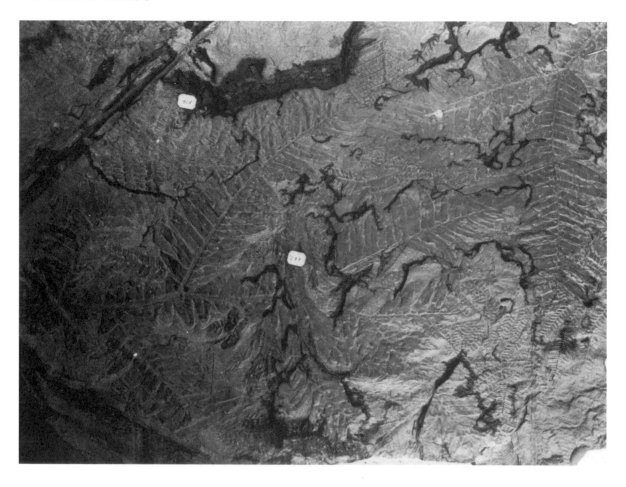

Figure 110 *Fossil leaves of the tree fern* Neuropteris, *Fayette County, Pennsylvania.*

(Photo by E. B. Hardin, courtesy USGS)

carbonized plant material is commonly found between easily separated sediment layers (Fig. 111). Animals were also buried in the great ancient coal swamps, where their bones were preserved and fossilized.

FOSSIL FUELS

The Carboniferous and Permian had the highest organic burial rates of any period in Earth history. Extensive forests and swamps grew successively on top of each other and continued to add to thick deposits of peat, which were buried under layers of sediment. The weight of the overlying strata and heat from the planet's interior reduced the peat to about 5 percent of its original volume and metamorphosed it into lignite as well as bituminous and anthracite coal.

The world's coal reserves far exceed all other fossil fuels combined. They are sufficient to support large increases in consumption well into this century.

The amount of economically recoverable coal is upward of 1 trillion tons. The United States holds substantial reserves of coal (Fig. 112), which remain practically untouched. Since coal is the cheapest and most abundant fossil fuel, it will be a favorable alternate source of energy to replace petroleum when reserves run low. However, because coal produces more pollution than other fossil fuels, new technology will be required to clean up coal-fired plants.

Paleozoic sediments hold a large portion of the world's oil reserves, indicating a high degree of marine organic productivity during this time. The formation of oil and gas requires special geologic conditions. These include a sedimentary source of organic material, a porous rock to serve as a reservoir, and a confining structure to act as a trap. The source material is organic carbon in fine-grained, carbon-rich sediments. Porous and permeable sedimentary rocks such as sandstones and limestones serve as reservoirs. Geologic structures created by folding or faulting of sedimentary layers trap or pool the oil and gas.

Most organic material that produces petroleum originates from microscopic organisms that originated primarily in the surface waters of the ocean and were concentrated in fine particulate matter on the seafloor. For organic material to become petroleum, either the rate of accumulation must be high

Figure 111 Fossilized plants of the upper Potsville series, Washington County, Arkansas.

(Photo by E. B. Hardin, courtesy USGS)

Figure 112 *Open pit coal mining at the West Decker mine, Montana.*

(Photo by P. F. Narten, courtesy USGS)

or the oxygen level in the bottom water must be low so the material does not oxidize before burial under thick sedimentary layers. This is because oxidation causes decay, which destroys organic material. Therefore, areas with high rates of accumulation of sediments rich in organic material are the most favorable sites for the formation of oil-bearing rock.

After deep burial in a sedimentary basin, high temperatures and pressures generated in Earth's interior chemically alter the organic material into hydrocarbons. If the hydrocarbons are overcooked, however, natural gas results. Oil is often associated with thick beds of salt. Because salt is lighter than the overlying sediments, it rises toward the surface, creating salt domes that help trap the oil.

Hydrocarbon volatiles (fluids and gases) along with seawater locked up in the sediments migrated upward through permeable rock layers and accumulated in traps formed by sedimentary structures that provide a barrier to further migration. In the absence of a cap rock, the volatiles continue rising to the surface and escape. Also, much petroleum has been lost by the destruction of the reservoir by uplift and erosion of the confining structure. Several tens of millions to a few hundred million years are required to process organic material into oil, mainly depending on the temperature and pressure conditions with the sedimentary basin.

CARBONIFEROUS GLACIATION

Terrestrial flora first appearing some 450 million years ago was plentiful and varied during the Carboniferous. Forests of seed ferns and true trees quickly spread across Gondwana and Laurasia in the early Carboniferous. Primitive amphibians inhabited the swampy forests. These forests were buzzing with hundreds of species of insects, including large cockroaches and giant dragonflies. When the climate grew colder and widespread glaciation enveloped the southern continents at the end of the period, the first reptiles emerged. They began displacing the amphibians as the dominant land vertebrates.

During the latter part of the Carboniferous around 290 million years ago, Gondwana was in the south polar regions. Glacial centers expanded across the southern continents, as evidenced by glacial deposits of tillites along with striations in ancient rocks. Rocks heavily grooved by the advancing glaciers show lines of ice flow away from the equator and toward the poles. This would not have been possible if the continents were situated where they are today. Furthermore, the ice would have had to flow from the sea onto the land in many areas, which is highly unlikely. Instead, the southern continents drifted en masse over the South Pole, and huge ice sheets crossed the present continental boundaries.

Glacial deposits were interbedded with marine sediments. Deposits of boulders distorted the finer sediment in which they lie. This indicates they fell onto the ocean's bottom muds from rafts of ice. Apparently, floating ice sheets extended outward from the land like they do today at Antarctica's huge ice shelves (Fig. 113). As the icebergs drifted away from the ice sheet and melted, debris embedded in the ice dropped into the sediment on the ocean floor over a wide area.

Strange, out-of-place boulders called erratics, which were composed of rock types not found elsewhere on one continent, matched rocks on the opposing continent. The glacial deposits were overlain by thick sequences of terrestrial sediments. In turn, these were covered by massive outpourings of basalt lava flows. Overlying these volcanic rocks were coal beds containing similar fossilized plant material.

Glacial deposits in present-day equatorial areas show that in the past these regions were much colder. Fossil coral reefs and coal deposits in the north polar regions suggest a tropical climate existed there at one time. More-

Figure 113 *Ice shelves of Daniell Peninsula, Victoria Land, Antarctica.*

(Photo by W. B. Hamilton, courtesy USGS)

over, salt deposits in the Arctic regions indicate an ancient desert climate. Either the climate in the past altered dramatically or the continents shifted positions with respect to the equator.

During the early part of the glacial epoch, the maximum glacial effects were in South America and South Africa. Later, as the continent drifted southward, the chief glacial centers moved to Australia and Antarctica, clearly showing that the southern continents that comprised Gondwana hovered over the South Pole. Continents residing near the poles is often the cause of extended periods of glaciation. This is because land at higher latitudes has a high albedo (reflective quality) and a low heat capacity, encouraging the accumulation of ice.

Ice sheets covered large portions of east-central South America, South Africa, India, Australia, and Antarctica. In Australia, marine sediments were interbedded with glacial deposits. Tillites were separated by seams of coal. These indicate that periods of glaciation were interspersed with warm interglacial spells, when extensive forests grew. In South Africa, the Karroo Series, comprising a sequence of late Paleozoic tillites and coal beds, extended over an area of several thousand square miles and reached a total thickness of 20,000 feet. Among the coal beds, the best in Africa, are fossil *Glossopteris* leaves. Their existence on the southern continents is among the best evidence for the theory of continental drift.

The glaciation was probably caused by the loss of atmospheric carbon dioxide. The burial of carbon dioxide in the crust might have been the key to the onset of all major ice ages since life evolved on Earth. Substantial carbon dioxide repositories during the latter part of the Paleozoic were the great forests that spread over the land. Plants invaded the land and extended to all parts of the world beginning about 450 million years ago. Lush forests that grew during the Carboniferous stored large quantities of carbon in their woody tissues. Burial under layers of sediment compacted the vegetative matter and converted it into thick seams of coal (Fig. 114). The reduction of the carbon dioxide content in the atmosphere severely weakened the greenhouse effect, causing the climate to cool.

The continental margins became less extensive and narrower, confining marine habitats to nearshore regions. This might have influenced the great extinction at the end of the Paleozoic. No major extinction event occurred during the widespread Carboniferous glaciation about 330 million years ago,

Figure 114 *A thick coal bed at Little Powder River coal field, Montana.*

(Photo by C. E. Dobbin, courtesy USGS)

Figure 115 *The supercontinent Pangaea 250 million years ago.*

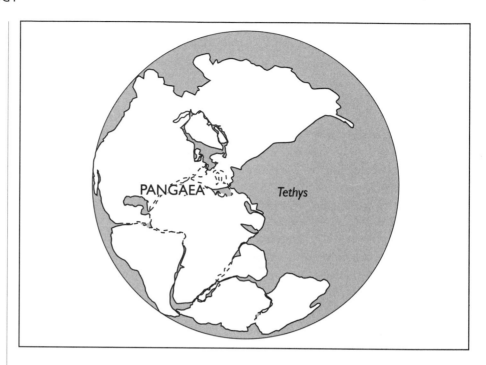

however. The relatively low extinction rates were probably due to a limited number of extinction-prone species following the late Devonian extinction.

When the glaciers departed, the first reptiles emerged to displace the amphibians as the dominant land vertebrates. The climate of the tropics became more arid, and the swamplands began to disappear. Land once covered with great coal swamps began to dry out as the climate grew colder.

PANGAEA

Between 360 and 270 million years ago, Gondwana and Laurasia converged into the supercontinent Pangaea (Fig. 115), whose name comes from Greek meaning "all lands." The massive continent had an area of about 80 million square miles and covered 40 percent of Earth's surface. It straddled the equator and extended practically from pole to pole, with an almost equal amount of land in both hemispheres. Today, in contrast, two-thirds of the continents lie north of the equator. A single great ocean called Panthalassa stretched uninterrupted across the planet, with all the continents huddling to one side. Over the ensuing periods, smaller parcels of land continued to collide with the supercontinent until it reached its peak size at the end of the Triassic about 210 million years ago.

The continental collisions crumpled the crust and pushed up huge masses of rocks into several mountain belts throughout many parts of the world (Fig. 116). Volcanic eruptions were extensive due to frequent continental collisions. During times of highly active continental movements, volcanic activity increases. This occurs especially at midocean spreading ridges where new oceanic crust is created and at subduction zones where old oceanic crust is destroyed. The amount of volcanism also affects the rate of mountain building and the composition of the atmosphere, which greatly influences the climate.

When Gondwana linked with Laurasia to form Pangaea, the collision raised the Appalachian and Ouachita Mountains. Simultaneously, Laurasia connected with Siberia, thrusting up the Ural Mountains. The continued clashing of island arcs with North America resulted in an episode of mountain building in Nevada called the Sonoma orogeny, which coincided with the assembly of Pangaea 250 million years ago.

North America sat astride the equator, which ran from northern Mexico to eastern Canada, during the Carboniferous 300 million years ago. It is believed to have had a wet, tropical climate. A great river system, rivaling today's Amazon, ran down from the ancient Appalachian Mountains. It flowed through valleys 20 miles wide while draining much of the continent. One river system, which drained an area comparable to that of the modern Mississippi River, had its headwaters in Canada's Maritime Provinces. It flowed southward for 2,300 miles, finally emptying near the present-day coastal plain of Alabama and Mississippi. A second river system began in south-central

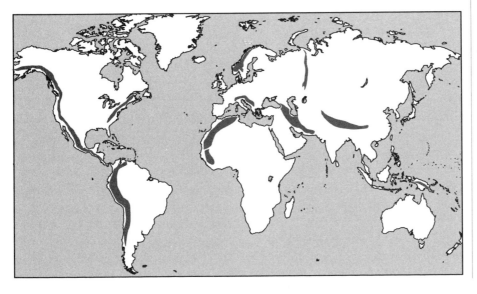

Figure 116 Major mountain ranges resulting from continental collisions.

Quebec and stretched all the way to Arkansas. The rivers reached their maximum in the early Carboniferous, during times of low sea levels. However, when sea levels rose, the rivers flooded the land and became vast swamps. Huge buildups of peat, which later became coal, account for the vast coal deposits in many areas of the eastern United States.

The sediments in the Tethys Sea separating Gondwana and Laurasia buckled and uplifted into various mountain belts, including the ancestral Hercynian Mountains of southern Europe. As the continents rose higher and the ocean basins dropped lower, the land became dryer and the climate grew colder, especially in the southernmost lands, which were covered with glacial ice. All known episodes of glaciation occurred during times of lowered sea levels. The changes in the shapes of the ocean basins greatly influenced the course of ocean currents, which in turn had a pronounced effect on the climate.

The closing of the Tethys Sea eliminated a major barrier to the migration of species from one continent to another, and they dispersed to every part of the world (Fig. 117). When all continents combined into Pangaea, plant and animal life witnessed a great diversity in the ocean as well as on land. The formation of Pangaea marked a major turning point in evolution of life, during which the reptiles emerged as the dominant species, conquering land, sea, and sky.

A continuous shallow-water margin ran around the entire perimeter of Pangaea. As a result, no major physical barriers hampered the dispersal of marine life. Furthermore, the seas were largely confined to the ocean basins, leaving the continental shelves mostly exposed. The continental margins were less extensive and narrower, confining marine habitats to the nearshore regions. Consequently, habitat areas for shallow-water marine organisms were very limited, which accounted for the low species diversity. As a result, marine biotas (flora and fauna) were more widespread but contained comparatively fewer species.

In the northern latitudes, thick forests of primitive conifers, horsetails, and club mosses that grew as tall as 30 feet dominated the mountainous landscape. Much of the interior probably resembled a grassless version of the steppes of central Asia, where temperatures varied from very hot in summer to extremely cold in winter. Since grasses would not appear for well over 100 million years, the scrubby landscape was dotted with bamboolike horsetails and bushy clumps of extinct seed ferns that resembled present-day tree ferns.

Browsing on the seed ferns were herds of moschops, 16-foot reptiles with thick skulls adapted for butting during mating season, a tactic similar to the behavior of modern herd animals. They were probably preyed upon by packs of lycaenops (Fig. 118), which were reptiles with doglike bodies and

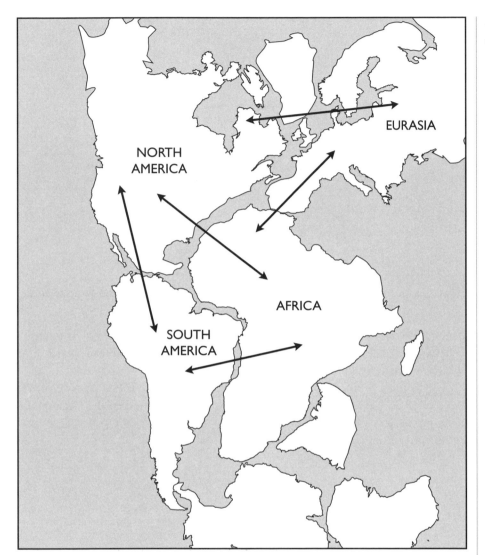

Figure 117 *The effect of geography on the migration of species.*

long canine teeth projecting from their mouths. Mammal-like reptiles called dicynodonts also had two caninelike tusks and fed on small animals along riverbanks. Small reptiles probably ate insects similar to the way modern lizards do.

The Pangaean climate was one of extremes. The northern and southern regions were colder than Siberia. The interior deserts were hotter than the Sahara and almost devoid of life. The massing of continents together created an overall climate that was hotter, drier, and more seasonal than at any other time in geologic history. These conditions might have led to the mass extinc-

Figure 118 *Lycaenops were reptiles that might have hunted in packs.*

tion of mostly terrestrial animals some 30 million years after most of Pangaea was assembled. Pangaea remained intact for another 40 million years, after which it rifted apart into the present continents.

After examining the amphibians of the Carboniferous coal forests, the next chapter explores the evolution of their cousins the reptiles during the Permian period.

9

PERMIAN REPTILES
THE AGE OF DESERT INHABITANTS

This chapter examines the evolution of the reptiles during the Permian period and covers the greatest mass extinction in Earth history. The Permian, from 280 to 250 million years ago, was named for a well-exposed sequence of marine rocks and terrestrial redbeds on the western side of the Ural Mountains in the Russian district of Perm. Rocks of Permian age are distinct in western North America, particularly Texas, Nevada, and Utah (Fig. 119). Important reserves of oil and natural gas reside in the Permian Basin of Texas and Oklahoma. Extensive coal deposits of Permian age exist in Asia, Africa, Australia, and North America.

During the Permian, all major continents were combined into the supercontinent Pangaea, where widespread mountain building and extensive volcanism were prevalent. The interior of Pangaea was largely desert, causing the decline of the amphibians in favor of the reptiles. At the end of the Permian, perhaps the greatest extinction Earth has ever known eliminated more than 95 percent of all species, paving the way for the ascension of the dinosaurs.

Figure 119 *Fremont Canyon with the Henry Mountains in the background, Wayne and Garfield Counties, Utah.*

(Photo by J. R. Stacy, courtesy USGS)

THE REPTILIAN ERA

The reptile age, which had its beginning in the Permian and lasted 200 million years, witnessed the evolution of some 20 orders of reptilian families. Amphibians, which were prominent in the Carboniferous, declined considerably in the Permian because of a preference for life in the water. When the Carboniferous swamps dried out and were largely replaced with deserts, the amphibians gave way to the reptiles, which were well adapted to drier climates. In the latter part of the Permian, the reptiles succeeded the amphibians and became the dominant land-dwelling animals of the Mesozoic era. The generally warm climate of the Mesozoic was advantageous to the reptiles and aided them in colonizing the land.

The increase in the number of reptilian fossil footprints in Carboniferous and Permian sediments shows the rise of the reptiles at the expense of the amphibians. The superiority of the reptiles was largely due to their more efficient mode of locomotion. Even at an early age, 290 million years ago, small reptiles were bipedaled and ran on two legs (Fig. 120), the fastest way to travel. They needed this speed not only to run down prey but also to escape a variety of meat-eating reptiles, including dimetrodon, a fierce Permian carnivore.

The most compelling evidence for bipedalism is that the length of the hind limbs are much longer than the forelimbs and would therefore make walking on four legs awkward. These were members of the diapsid group, one of the most primitive reptile lineages. It gave rise to dinosaurs, birds, and most living reptiles, including crocodiles, lizards, and snakes. The reptiles were also better suited to a full-time life on dry land. In contrast, the amphibians depended on a local source of water for moistening their skins and for reproduction.

The reptilian foot was a major improvement over that of the amphibian, with changes in the form of the digits, the addition of a thumblike fifth digit, and the appearance of claws. In some reptiles, the tracks narrowed and the stride lengthened. Others maintained a four-footed walking gait and ran

Figure 120 *The small plant eater camptosaur, an ancestor of many later dinosaurs, ran on its hind legs for speed and agility.*

reared up on their hind legs. Although most reptiles walked or ran on all fours, by the late Permian, some smaller reptiles often stood on their hind legs when they required speed and agility. The body pivoted at the hips, and a long tail counterbalanced the nearly erect trunk. This stance freed the forelimbs for attacking small prey and completing other useful tasks.

Reptiles have scales that retain the animal's bodily fluids. In contrast, amphibians have a permeable skin that must be moistened frequently. Another major advancement over the amphibians was the reptiles' mode of reproduction. Like fish, amphibians laid their eggs in water. After hatching, the young fended for themselves, often becoming prey for predators. The reptiles' eggs had hard, watertight shells so they could be laid onto dry land. Reptiles belong to a group known as amniotes, which also includes birds and mammals. Amniotes are animals with complex eggs that evolved from the amphibians. Reptile parents protected their young, which gave them a better chance of survival, contributing to reptiles' great success in populating the land.

Like fish and amphibians, reptiles are cold-blooded, a term that is misleading since they draw heat from the environment. Therefore, the blood of a reptile sunning on a rock can actually be warmer than that of a warm-blooded mammal. A high body temperature is as important to a reptile as it is to a mammal to achieve maximum metabolic efficiency. On cold mornings, reptiles are sluggish and vulnerable to predators. They bask in the sun until their bodies warm and their metabolism can operate at peak performance.

Reptiles require only about one-tenth the amount of food mammals need to survive because mammals use most of their calories to maintain a high body temperature. The total energy consumption of mammals is 10 to 30 times and the oxygen intake is about 20 times that of reptiles the same weight. Consequently, reptiles can live in deserts and other desolate places and flourish on small quantities of food that would quickly starve a mammal of the same size. The generally warm climate of the Mesozoic was very advantageous to the reptiles and aided them in colonizing the land. In contrast, the amphibians, which avoided direct sunlight and were relatively cold and slow moving, were at a disadvantage.

Many early reptiles evolved into some of the most bizarre creatures. Perhaps the strangest reptile that ever lived was *Tanystropheus,* dubbed the "giraffe-neck saurian." The animal measured as much as 15 feet from head to tail and is famous for its absurdly long neck, which was more than twice the length of the trunk. As it matured, its neck grew at a much faster rate than the rest of its body. Apparently, the reptile was aquatic because it could not possibly have supported the weight of such a lengthy neck while on land. *Tanystropheus* probably stretched its neck downward to scavenge bottom sediments for food.

The phytosaurs were large, heavily armored predatory reptiles with sharp teeth. They resembled crocodiles with their elongated snouts, short legs, and long tails but were not closely related to them. They evolved from the thecodonts, which also gave rise to the crocodiles and dinosaurs. They thrived in the late Triassic, evolving quite rapidly, but apparently did not survive beyond the end of the period.

Near the close of the Triassic, when reptiles were the leading form of animal life, occupying land, sea, and air, a remarkable reptilian group called the crocodilians appeared in the fossil record. The crocodilians originated in Gondwana. They comprised the alligators with a blunt head, the crocodiles with an elongated head, and the gavials with an extremely narrow head. Members of this group adapted to life on dry land, a semiaquatic life, or an entirely aquatic life with a sharklike tail, a streamlined head, and legs remolded into swimming paddles.

As with many land animals living in the Cretaceous around 100 million years ago, crocodiles grew to giant sizes as indicated by a 40-foot-long fossil found in southern Brazil. A fossil of a gaviallike monster from the lower Cretaceous in Niger, West Africa, measured about 35 feet long. A monstrous three-ton, 30-foot crocodile terrorized the Cretaceous swamps and might have preyed upon medium-sized dinosaurs.

The crocodilians diversified considerably over the past 200 million years. They spread to all parts of the world and adapted to a wide variety of habitats. Crocodile fossils found in the high latitudes of North America (Fig. 121) indicate a warm climate during the Mesozoic. They belong to the subclass Archosauria, which literally means "ruling reptiles," along with dinosaurs and pterosaurs. They are the only surviving members to escape the great dying at the end of the Cretaceous.

Reptiles that returned to the sea to compete with the fish for a plentiful food supply included the sea serpent–like plesiosaurs, the sea cow–like placodonts, and the dolphinlike mixosaurs. The pachycostasaurs, meaning "thick-ribbed lizard," were carnivorous marine reptiles 9 feet or more long. They resembled plesiosaurs, with thick, heavy ribs possibly used to house very large lungs and used for ballast in order to hunt at great depths.

The sharklike ichthyosaurs (Fig. 122), whose name is Greek for "fish lizard," were fast-swimming, shell-crushing marine predators that apparently preyed on ammonites. The ichthyosaur would first puncture the shell from the victim's blind side, causing it to fill with water and sink to the bottom, where the attack could be made head on. Puncture marks on fossil ammonite shells are spaced the same distance apart as fossil ichthyosaur teeth, suggesting these highly aggressive predators might have hastened the extinction of most ammonite species by the end of the Mesozoic.

The placodonts were a group of short, stout marine reptiles with large, flattened teeth. They probably fed primarily on bivalves and other mollusks. Several other reptilian species also went to sea, including lizards and turtles that were quite primitive. Many modern giant turtles are descendants of those

Figure 121 *The Chinle Formation, Uinta Range, Utah, where crocodile fossils are found.*

(Photo courtesy of the author)

marine reptiles. However, only the smallest turtles survived the extinction at the end of the Cretaceous.

Turtles are the closest living relatives of crocodiles, another extremely successful aquatic reptile. The crocodiles, which passed through the extinction practically unscathed, are closer to turtles than they are to lizards, snakes, and birds. Turtles, which are holdovers from an ancient group called anapsids that

Figure 122 *The ichthyosaur was an air-breathing reptile that returned to the sea.*

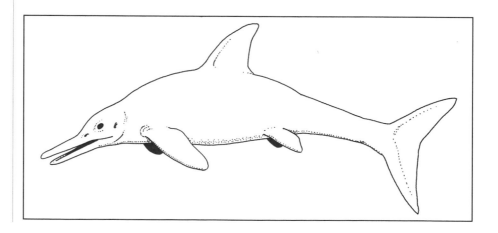

Figure 123 *Dimetrodon used the huge sail on its back to regulate its body temperature.*

lacked holes in the sides of their skulls, were once regarded as outsiders among modern reptiles because living reptiles and birds termed diapsids have two holes in the sides of their skulls.

MAMMAL-LIKE REPTILES

The mammal-like reptiles were animals in transition from reptiles to mammals. Fossilized bones of a mammal-like reptile with large down-pointing tusks called *Lystrosaurus* were discovered in the Transantarctic Range of Antarctica. Its presence provides strong evidence that the continent had once been joined with southern Africa and India. As an ancestor to mammals, lystrosaurus was the most common vertebrate on land and was found throughout Pangaea. Mammal-like reptiles called dicynodonts also had two caninelike tusks and fed on small animals along riverbanks.

The pelycosaurs were the first animals to depart from the basic reptilian stock some 300 million years ago. They were distinguished from other reptiles by their large size and varied diet. The earliest predators could kill relatively large prey, including other reptiles. A pelycosaur called dimetrodon (Fig. 123) obtained a length of about 11 feet. It had a large dorsal sail composed of webs of membrane well supplied with blood, stretching across bony protruding spines and probably used for temperature control. When the animal was cold, it turned its body broadside to the sun to absorb more sunlight. When the animal was hot, it sought out a shaded area or exposed itself to the wind. This appendage might have been a crude forerunner of the temperature control system in mammals.

As the climate warmed, the pelycosaurs lost their sails and perhaps gained some degree of internal thermal control. They thrived for about 50 million years. Then they gave way to their descendants, the mammal-like reptiles called the therapsids. The first therapsids retained many characteristics of the pelycosaurs, with legs better adapted for much higher running speeds. They ranged in size from as small as a mouse to as large as a hippopotamus.

The early members invaded the southern continents at the beginning of the Permian when those lands were recovering from the Carboniferous glaciation. This suggests the animals were sufficiently warm-blooded to withstand the cold. They had probably undergone some physiological adaptations to enable them to feed and travel through the snows of the cold winters. They were apparently too large to hibernate, as shown by the lack of growth rings in their bones similar to tree rings that mark alternating seasons of growth. The development of fur appeared in the more advanced therapsids as they migrated into colder climates. Therapsids might also have operated at lower body temperatures than most living mammals to conserve energy.

The family of mammal-like reptiles clearly shows a transition from reptile to mammal. Mammals evolved from the mammal-like reptiles over a period of more than 100 million years. During that time, the animals adapted so as to function better in a terrestrial environment. Teeth evolved from simple cones replaced repeatedly during the animal's lifetime to more complex shapes replaced only once upon maturity. However, the mammalian jaw and other parts of the skull still shared many similarities with reptiles.

Mammals are fully warm-blooded. The advantages of being warm-blooded are tremendous. A stable body temperature finely tuned to operate within a narrow thermal range provides a high rate of metabolism independent of the outside temperature. Therefore, the work output of legs, heart, and lungs increases enormously, giving mammals the ability to out perform and out endure reptiles. The principle of heat loss for large reptiles, by which a large body radiates more thermal energy than a small one, applies to large reptiles as well as mammals. In addition, mammals have a coat of insulation including an outer layer of fat and fur to prevent the escape of body heat during cold weather.

The therapsids appear to have reproduced like reptiles by laying eggs. They might have protected and incubated the eggs and fed their young. This, in turn, might have resulted in longer egg retention in the female and led to live births. The therapsids dominated animal life for more than 40 million years until the middle Triassic. Then for unknown reasons, they lost out to the dinosaurs. From then on, primitive mammals were relegated to the role of a shrewlike nocturnal hunter of insects until the dinosaurs finally became extinct.

THE APPALACHIAN OROGENY

Perhaps the most impressive landforms on Earth are mountain ranges, created by the forces of uplift and erosion. Paleozoic continental collisions crumpled the crust, pushing up huge masses of rocks into several mountain chains throughout many parts of the world. Mountains have complex internal structures formed by folding, faulting, volcanic activity, metamorphism, and igneous intrusion (Fig. 124).

The Appalachian Mountains, extending some 2,000 miles from central Alabama to Newfoundland, were upraised during continental collisions between North America, Eurasia, and Africa. This occurred in the late Paleozoic from about 350 million to 250 million years ago during the construction of Pangaea. The southern Appalachians were underlain with more than 10 miles of flat-lying sedimentary and metamorphic rocks, whereas the surface rocks were highly deformed by the collision.

This type of formation suggests that these mountains were the product of thrust faulting, involving crustal material carried horizontally for great distances. The sedimentary strata rode westward on top of Precambrian metamorphic rocks and folded over, buckling the crust into a series of ridges and

Figure 124 *Bear Butte near Sturgis, South Dakota, is an exposed granitic intrusion called a laccolith, showing outcrops of upturned sedimentary rocks around its base.*

(Photo by N. H. Darton, courtesy USGS)

Figure 125 *The Blue Ridge Mountains in the Appalachians, Avery County, North Carolina.*

(Photo by D. A. Brobst, courtesy USGS)

valleys (Fig. 125). The existence of sedimentary layers beneath the core of the Appalachians suggests that thrusting involving basement rocks is responsible for the formation of mountain belts, possibly since the process of plate tectonics began. The shoving and stacking of thrust sheets during continental collisions might also have been a major mechanism in the continued growth of the continents.

The Mauritanide Mountain Range in Western Africa is the counterpart (the other side) of the Appalachians. It is characterized by a series of belts running east to west that are similar in many respects to the Appalachian belts. The eastern parts of the range comprise sedimentary strata partially covered with metamorphic rocks that have overridden the sediments from the west along thrust faults. Older metamorpohic rocks resembling those of the southern Appalachians lie westward of this region, while a coastal plain of younger horizontal rocks covers the rest. Furthermore, a period of metamorphism and thrusting similar to the formation of the Appalachians occurred before the opening of the Atlantic. In this respect, the two mountain ranges are practically mirror images of each other.

This episode of mountain building also uplifted the Hercynian Mountains in Europe, which extended from England to Ireland and continued through France and Germany. The folding and faulting were accompanied by large-scale igneous activity in England and on the European continent. The Ural Mountains were similarly formed by a collision between the Siberian and Russian shields. The Transantarctic Range, comprising great belts of folded rocks, formed when two plates came together into the continent of Antarctica. Prior to the end of the Permian, the younger parts of West Antarctica had not yet formed, and only East Antarctica was present.

LATE PALEOZOIC GLACIATION

The continents of Africa, South America, Australia, India, and Antarctica were glaciated in the late Paleozoic, around 270 million years ago, as evidenced by glacial deposits and striations in ancient rocks. The continents were joined in such a manner that ice sheets moved across a single landmass, radiating outward from a glacial center over the South Pole.

The late Paleozoic was a period of extensive mountain building. Massive chunks of crust were raised to higher elevations, where glaciers are nurtured in the cold, thin air. Glaciers might have also formed at lower latitudes when lands were elevated during continental collisions. When Gondwana and Laurasia combined into Pangaea, the continental collisions crumpled the crust and pushed up huge blocks into several mountain chains throughout many parts of the world. With land at high elevations, temperature decreases and precipitation increases, maintaining glaciers in the high altitudes. Evidence for widespread glaciation include beds of Permian tillites found on nearly every continent (Fig. 126).

Besides folded mountain belts, volcanoes were prevalent. Unusually long periods of volcanic activity block out the Sun with clouds of volcanic dust and

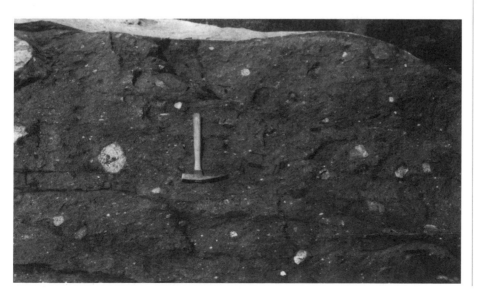

Figure 126 Permian tillite in Roxbury conglomerate, Hyde Park, Suffolk County, Massachusetts.

(Photo by W. C. Alden, courtesy USGS)

TABLE 7 HISTORY OF THE DEEP CIRCULATION
OF THE OCEAN

Age (millions of years ago)	Event
3	An Ice Age overwhelms the Northern Hemisphere.
3–5	Arctic glaciation begins.
15	The Drake Passage is open; the circum-Antarctic current is formed. Major sea ice forms around Antarctica, which is glaciated, making it the first major glaciation of the modern Ice Age. The Antarctic bottom water forms. The snow limit rises.
25	The Drake Passage between South America and Antarctica begins to open.
25–35	A stable situation exists with possible partial circulation around Antarctica. The equatorial circulation is interrupted between the Mediterranean Sea and the Far East.
35–40	The equatorial seaway begins to close. There is a sharp cooling of the surface and of the deep water in the south. The Antarctic glaciers reach the sea with glacial debris in the sea. The seaway between Australia and Antarctica opens. Cooler bottom water flows north and flushes the ocean. The snow limit drops sharply.
> 50	The ocean could flow freely around the world at the equator. Rather uniform climate and warm ocean occur even near the poles. Deep water in the ocean is much warmer than it is today. Only alpine glaciers exist on Antarctica.

gases, thereby significantly lowering surface temperatures. As the continents rose higher, the ocean basins dropped lower. The change in shape of the ocean basins greatly affected the course of ocean currents (Table 7), which, in turn, had a profound effect on the climate.

All known episodes of glaciation occurred at times when sea levels should have been low, although not all mass extinctions were associated with lowered sea levels. The recession of the seas made continental margins less extensive and narrower, confining marine habitats to nearshore areas. Such an occurrence might have had a major influence on the great extinction at the end of the Paleozoic era. During this time, land once covered with great coal swamps completely dried out as the climate grew colder, culminating in the deaths of large numbers of species.

Ocean temperatures also remained cool for a considerable time following the late Permian ice age. Episodes of climatic cooling are detrimental to species that do not adapt to the new, colder conditions or migrate to warmer refuge. Marine invertebrates that escaped extinction lived in a narrow margin near the equator. Corals, which live only in warm, shallow water, were particularly hard hit as evidenced by the lack of coral reefs in the early Mesozoic. When the great glaciers melted and the seas began to warm to their preglacial conditions, reef building intensified, forming thick deposits of limestone that were laid down by prolific, lime-secreting organisms.

MASS EXTINCTION

Throughout Earth history, massive numbers of species have vanished in several short periods (Fig. 127). During geologically brief intervals of perhaps a few million years, mass extinctions in the ocean have eliminated half or more of the existing families of plants and animals. Devastation of this magnitude resulted from radical global changes in the environment. Drastic changes in environmental limiting factors, including temperature and living space on the ocean floor, determined the distribution and abundance of species in the sea.

Many episodes of extinction coincided with periods of glaciation, and global cooling had a major effect on life. The living space of warmth-loving

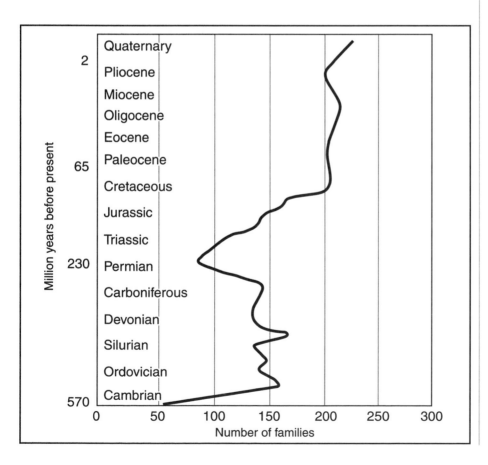

Figure 127 *The number of families through time. The large drop at the end of the Permian indicates a major mass extinction.*

171

species was restricted to narrow areas around the tropics. Species trapped in confined waterways that were unable to move to warmer seas were particularly hard hit. Furthermore, the accumulation of glacial ice in the polar regions lowered sea levels, thereby reducing shallow water shelf areas. This limited the amount of habitat and consequently the number of species.

Ocean temperature is by far the most important factor limiting the geographic distribution of marine species. Climatic cooling is the primary culprit behind most extinctions in the sea. Species unable to migrate to warmer regions or adapt to colder conditions are usually the most adversely affected. This is especially true for tropical faunas that can tolerate only a narrow range of temperatures and have nowhere to migrate. Since lowered temperatures also slow the rate of chemical reactions, biological activity during a major glacial event should function at a lower energy state, which could affect the rate of evolution and species diversity.

The greatest extinction event took place when the Permian ended 250 million years ago. The extinction was particularly devastating to Permian marine fauna (Fig. 128). Half the families of marine organisms, including more than 95 percent of all known species, abruptly disappeared. On land, more than 70 percent of the vertebrates died out. So many plants succumbed to extinction that fungi briefly ruled the continents. Two distinct die outs occurred in the space of 1 to 2 million years. About 70 percent of the species became extinct during the first event, and 80 percent of the remaining species died out in the second episode. The most pronounced loss of species took place between 252 and 251 million years ago. The final extinction pulse might have lasted less than 1 million years. In effect, the extinction left the world almost as devoid of species at the end of Paleozoic as at the beginning.

Groups attached to the seafloor and filtered organic material from seawater for nutrients suffered the greatest extinction. Corals, which require warm, shallow water for survival, were hardest hit by the extinction. They were followed by brachiopods, bryozoans, echinoderms, ammonoids, foraminifera, and the last remaining trilobites. Another major group of animals that disappeared were the fusulinids, which populated the shallow seas for about 80 million years, during which their shells accumulated into vast deposits of limestone. They were large, complex protozoans resembling grains of wheat. They ranged from microscopic size up to 3 inches in length. Planktonic plants also died out, devastating the base of the marine food web, upon which other species depended for their survival.

The more mobile creatures, such as bivalves, gastropods, and crabs, escaped the extinction relatively unharmed. Those organisms, especially the shelled types, because they could buffer their internal organs from changes in ocean chemistry, were less likely to be wiped out. They were better able to

rebound after a mass extinction than their more sensitive neighbors. Nevertheless, both groups declined considerably. However, the unbuffered organisms were the hardest hit, losing 90 percent of their genera compared with 50 percent for the buffered group.

Surviving crinoids and brachiopods, which were highly prolific in the Paleozoic, were relegated to minor roles during the succeeding Mesozoic era. The spiny brachiopods, which were plentiful in the late Paleozoic seas, vanished entirely without leaving any descendants. The trilobites, which were extremely successful during most of the Paleozoic, suffered final extinction when the era ended. A variety of other crustaceans, including shrimps, crabs, crayfish, and lobsters, occupied habitats vacated by the trilobites. The sharks, regaining ground lost by the great extinction, continued to become the successful predators they are today.

On land, some 70 percent of the amphibian families and about 80 percent of the reptilian families also disappeared. Even the insects did not escape the carnage. Nearly one-third of their orders living in the Permian did not survive, marking the only mass extinction insects have ever undergone. The loss of plant life might have contributed to the disappearance of insects that fed on the vegetation. Following the extinction, insects shifted from a variety of dragonflylike groups, whose wings were fixed in the flight position, to forms that could fold their wings over their bodies, possibly as a means of protection.

The extinction followed on the heels of a late Permian glaciation, when thick sheets of ice blanketed much of the planet, significantly lowering ocean temperatures. As a further blow, one of the largest volcanic outpourings known on Earth covered northern Siberia in thick layers of basalt, causing considerable changes in the environment during a 1-million-year period. The eruptions removed much of Earth's oxygen and replaced it with choking sulfurous and carbon dioxide gases. Fossil evidence suggests that the Permian extinctions began gradually, culminating with a more rapid pulse at the end possibly due to environmental chaos.

The interior seas retreated from the continents as an abundance of terrestrial redbeds and large deposits of gypsum and salt were laid down. Extensive mountain building raised massive chunks of crust. Land at higher elevations accumulated snow and ice, which accumulated into thick glaciers. The glaciers, in turn, reflected sunlight, causing surface temperatures to drop. The most important factor limiting the geographic distribution of marine animal species is water temperature. Corals, which require warm, shallow water, were affected the worst as evidenced by the lack of coral reefs in the early Triassic.

Whatever were the agents of biological stress—climatic changes, shifts in ocean currents, shallow seas, or disruptions in food chains—the ability of the biosphere to resist them varied in different parts of the world. However, one pattern of mass extinctions occurred very consistently. Although each event typically affected different suites of organisms, tropical biotas, which contain the highest number of species, were almost always the hardest hit.

When all continents had converged into Pangaea by the end of the Permian around 250 million years ago, the change in geography spurred a great diversity of plant and animal life on land and in the sea. The formation of Pangaea marked a major turning point in the evolution of life, during which the reptiles emerged as the dominant species of the world.

The Pangaean climate appears to have been equable and warm throughout most of the year. However, much of the interior of Pangaea was desert, whose temperatures fluctuated wildly from season to season. It had scorching hot summers and freezing cold winters. These climate conditions might have contributed to the widespread extinction of land-based species during the late

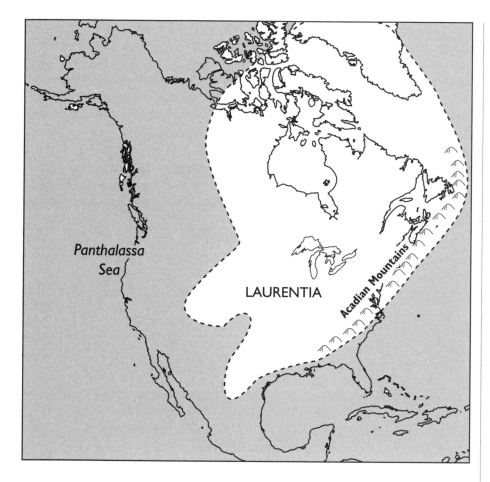

Paleozoic. It also explains why the reptiles, which adapt readily to hot, dry climates, replaced the amphibians as the dominant land animals.

The sea level lowered dramatically during the formation of Pangaea, which drained the interiors of the continents (Fig. 129). The drop in sea level caused the inland seas to retreat, producing a continuous, narrow continental margin around the supercontinent Pangaea. This, in turn, reduced the amount of shoreline, confining marine habitats to nearshore environments. This had a major influence on the extinction of marine species. Moreover, unstable nearshore conditions resulted in an unreliable food supply. Many marine species unable to cope with the limited living space and food supply died out in tragically large numbers.

The extinction was particularly devastating to Permian marine fauna. Half the families of aquatic organisms, 75 percent of the amphibian families, and over 80 percent of the reptilian families, representing more than 95 percent of all known species, abruptly disappeared. In effect, the extinction left the world practically devoid of species, thereby paving the way for the ascension of entirely new life-forms.

In the aftermath of the Permian extinction, many once minor groups, including active predatory relatives of modern fish, squids, snails, and crabs, were able to expand. For example, sea urchins, which were relatively uncommon during the Permian, are widespread in today's oceans. Since the great Permian catastrophe, eight significant extinction events have occurred. These defined the boundaries of the geologic time scale, with many of the strongest peaks coinciding with the demarcations between geologic periods.

After discussing the reptiles in the Permian, the next chapter will examine the evolution of the dinosaurs and their geologic environment in the Triassic period.

10

TRIASSIC DINOSAURS
THE AGE OF BIG BEASTS

T his chapter examines the evolution of the dinosaurs and of the continents during the Triassic period. The Triassic, from 250 to 210 million years ago, marks the beginning of the Mesozoic era. The period was named for a sequence of redbed and limestone strata in central Germany. All continents joined into a single great continent called Pangaea, which supported a distinctive suite of terrestrial plants and animals. In North America, continental sediments and redbeds add to the rugged beauty of Wyoming, Colorado, and Utah (Fig. 130). At the end of the period, Pangaea began rifting apart into the present-day continents, and massive amounts of basalt spilled onto the landscape.

During the late Triassic, large families of terrestrial animals died off in record numbers. The extinction spanned a period of less than a million years and was responsible for killing nearly half the reptile families. The dying out of species forever changed the character of life on Earth and initiated the rise of the dinosaurs (Fig. 131), one of biology's greatest success stories.

THE DINOSAUR ERA

At the beginning of the Mesozoic, the continents consolidated into a super-continent; at midpoint, they began to rift apart; and at the end, they were well

Figure 130 *Triassic and Jurassic beds at Silver Falls Creek, Circle Cliffs region, Garfield County, Utah.*

(Photo by R. G. Luedke, courtesy USGS)

on their way to their present-day locations. The breakup of Pangaea created three new bodies of water, including the Atlantic, Arctic, and Indian Oceans. The climate was exceptionally mild for an unusually long time. One particular group of animals that excelled during these extraordinary conditions were the reptiles. Besides conquering the land, some species went to sea and others took to the air, occupying nearly every corner of the globe. Forests were also extensive and covered large parts of the world with lush vegetation. In Arizona's Petrified Forest lie the fossilized remains of primitive Triassic-age conifers (Fig. 132) that once flourished in the upland regions.

Early in the Triassic, Earth was recovering from a major ice age and an extinction event that took the lives of more than 95 percent of all species. Thus, the start of the Mesozoic marked a rebirth of life, and 450 new families of plants and animals came into existence. However, instead of inventing entirely new body plans as happened during the Cambrian explosion at the beginning of the Paleozoic, species developed new variations on already established themes. Therefore, fewer experimental organisms arose, and many lines of today's species evolved. Several major groups of terrestrial vertebrates made their debut. These comprised the direct ancestors of modern animals, including reptiles, mammals, and perhaps the predecessors of birds. (The true birds did not appear in the fossil record for another 50 million years.)

The amphibians declined substantially when the great swamps began to dry out toward the end of the Paleozoic. Populations of amphibians continued to fall during the Mesozoic. All large, flat-headed species became extinct early in the Triassic. The group thereafter was represented by more modern forms. Although the amphibians did not achieve complete dominion over the land, their descendants the reptiles became the undisputed rulers of the world.

The reptiles were the leading form of animal life in the early Triassic. The unusually warm climate of the Mesozoic contributed substantially to the success of the reptiles. Not all reptiles were small creatures, however; some grew upward of 16 feet long. The reptiles were better suited for living continuously

Figure 131 *Dinosaurs were one of biology's greatest success stories.*

Figure 132 *Petrified trees at the Petrified Forest National Monument, Apache County, Arizona.*

(Photo by N. H. Darton, courtesy USGS)

on dry land. They successfully inhabited desert areas in the interiors of continents and other desolate places where they could flourish on small amounts of food. Many reptiles walked four footed but reared up on their hind legs when running down prey, thus freeing the forelimbs for attacking other animals.

The earliest mammals saw their beginning in the Triassic. At the start of the dinosaur era, mammals lived alongside the great beasts. At this time, mammals were small, ratlike creatures of little consequence. However, in the Mesozoic mammals underwent some remarkable adaptations that in many ways dwarfed those of their larger reptilian neighbors. Mammals developed specialized teeth, highly sensitive hearing and sight, and enlarged brains that would later help them outlive the dinosaurs.

Teeth are often the only remains of tiny, extinct mammals. However, they reveal much about the animals such as their diet over the past 220 million years. Chief among these early mammals were the multituberculates, which first appeared in the upper Triassic 210 million years ago. Unfortunately, after having survived the dinosaur extinction at the end of the Cretaceous, they all died out in the Eocene more than 30 million years ago.

Insects can easily claim the title of the world's most prosperous creatures. Ever since animals left the oceans and took up residence on dry land, insects and their arthropod relatives have ruled the planet. Paleontologists discovered a missing chapter in insect evolution in an extraordinarily rich fossil quarry in southern Virginia, near the North Carolina border. The site yielded well-preserved insects dating to the end of the Triassic 210 million years ago, including the discovery of the oldest records of some types of insects. Insects are hard to fossilize because their delicate bodies tend to disintegrate on burial. Later insects were preserved in amber, which is altered tree sap that entombs the animals and enables them to withstand the rigors of time.

The oldest dinosaurs originated on the southern continent Gondwana when the last glaciers of the great Permian ice age were departing and the region was still recovering from the cold conditions. The dinosaurs descended from the thecodonts (Fig. 133), the apparent common ancestors of crocodiles and birds, their distant living relatives. The earliest thecodonts were small- to medium-sized predators that lived during the Permian-Triassic transition. One group of thecodonts took to the water and became large fish eaters. They included the phytosaurs, which died out in the Triassic, and the crocodilians, which remain successful today.

Pterosaurs were also descendants of the Triassic thecodonts. The appearance of featherlike scales ostensibly used for insulation suggests that thecodonts were also the ancestors of birds. The protofeathers helped trap body heat or served as a colorful display for attracting mates as with modern birds. By the end of the Triassic, the dinosaurs replaced the thecodonts as the dominant terrestrial vertebrates.

Every dinosaur species could trace its origin back to a single common ancestor called eoraptor, meaning "early hunter." It evolved from predatory reptiles in the lower Triassic about 240 million years ago. Early in dinosaur

Figure 133 The dinosaurs descended from the thecodonts.

evolution, the animals had to compete with huge crocodiles, flying reptiles called pterodactyls, and other fierce reptiles. In the middle Triassic about 230 million years ago, when mammal-like reptiles dominated the land, dinosaurs represented only a minor percentage of all animals. Several reptile species living at the time of the early dinosaurs still far outweighed them.

In just 10 million years, however, dinosaurs rapidly became the dominant species. They evolved from moderate-sized animals less than 20 feet long to the giants for which they are famous. Dinosaurs arose to become the dominant terrestrial species for the next 150 million years. By the end of the Jurassic, dinosaurs were the largest predators ever to roam Earth. Their fossils are found in relative abundance in Triassic to Cretaceous formations (Fig. 134).

The dinosaurs ventured to all major continents. Their distribution throughout the world is strong evidence for continental drift (Table 8). At the time the dinosaurs came into existence, all continents were assembled into the supercontinent Pangaea. Early in the Jurassic, it began to rift apart, and the continents drifted toward their present locations. Except for a few temporary land bridges, the oceans that filled the rifts between the newly formed continents provided a barrier to further dinosaur migration. At this time, almost identical species lived in North America, Europe, and Africa.

Figure 134 *The Carmel Formation, Uinta Range, Utah, where dinosaur fossils are found.*

(Photo courtesy of the author)

TABLE 8 CONTINENTAL DRIFT

Geologic Division (millions of years)		Gondwana	Laurasia
Quaternary	3		Opening of the Gulf of California
Pliocene	11	Begin spreading near Galapagos Islands	Change spreading directions in eastern Pacific
Miocene	26	Opening of the Gulf of Aden Opening of the Red Sea	Birth of Iceland
Oligocene	37	Collision of India with Eurasia	Begin spreading in the Arctic Basin
Eocene	54		Separation of Greenland from Norway
Paleocene	65	Separation of Australia from Antarctica	
		Separation of New Zealand from Antarctica	Opening of the Labrador Sea Opening of the Bay of Biscay
		Separation of Africa from Madagascar and South America	Major rifting of North America from Eurasia
Cretaceous	135	Separation of Africa from India, Australia, New Zealand, and Antarctica	
Jurassic	180		
			Begin separation of North America from Africa
Triassic	250		

The success of the dinosaurs is exemplified by their extensive range. They occupied a wide variety of habitats and dominated all other forms of land-dwelling animals. Indeed, had the dinosaurs not become extinct, mammals would never have achieved dominance over Earth and humans would not have come into existence. This is because the dinosaurs would have continued to suppress further advancement of the mammals, which would have remained small, nocturnal creatures, keeping out from underfoot of the dinosaurs.

Dinosaurs are classified as either sauropods or carnosaurs. Sauropods were long-necked herbivores. Carnosaurs were bipedal carnivores that possibly hunted sauropods in packs. Camptosaur (Fig. 135), ancestor of many later dinosaur species, was a herbivore up to 25 feet long. Not all dinosaurs were gigantic, however. Many were about the size of most modern mammals. *Protoceratops* and ankylosaurs were no larger than the biggest terrestrial mammals

Figure 135 *The small plant eater camptosaur was an ancestor of many later dinosaurs.*

(Photo courtesy National Park Service)

living today. They were very common and ranged over wide spaces like modern-day sheep. The smallest known dinosaur footprints are only about the size of a penny. The smaller dinosaurs had hollow bones similar to those of birds. Some had long, slender hind legs; long, delicate forelimbs; and a long neck. If not for a lengthy tail, their skeletons would have closely resembled those of modern ostriches.

Many early small dinosaurs reared up on their hind legs and were among the first animals to establish a successful permanent two-legged stance. They are descendants of the diapsid reptiles, which ran on two legs, the fastest way to travel. The diapsids were one of the most primitive reptile lineages. They gave rise to dinosaurs, birds, and most living reptiles, including crocodiles, lizards, and snakes. Bipedalism increased speed and agility. It also freed the forelimbs for foraging and other tasks. The back legs and hips supported the entire weight of the animal, while a large tail counterbalanced the upper portions of the body. The dinosaur walked much like birds. Therefore, dinosaurs are classified either as ornithischians with a birdlike pelvis or as saurischians with a lizardlike pelvis.

The ornithischians probably arose from the same group of thecodont reptiles that gave rise to crocodiles and birds. Indeed, birds are the only living relatives of dinosaurs. The skeletons of many small dinosaurs resemble those of birds (Fig. 136). Many large bipedaled dinosaurs assumed similar appearances simply because of the need to balance themselves on two feet, which required a huge tail and small forearms. A two-legged sprint would also have been the fastest way to travel to run down prey.

Some dinosaur species might have acquired a certain degree of body temperature control similar to mammals and birds. When the dinosaur age began, the climates of southern Africa and the tip of South America where the early dinosaurs roamed experienced cold winters, during which large, cold-blooded animals could not have survived without migrating to warmer regions. The stamina needed for such long-distance migration would have required sustained energy levels that only warm-blooded bodies could provide. Warm-blooded animals mature more rapidly than cold-blooded reptiles, which continue growing steadily until death.

An analysis of the bones of dinosaurs, crocodiles, and birds, all of which had a common ancestor, shows a similarity between bird and dinosaur bones—another sign of possible warm-bloodedness. The complex social behavior of dinosaurs that requires a high rate of metabolism appears to be an evolutionary advancement that results from being warm-blooded. Yet at the end of the Cretaceous, when the climate is thought to have grown colder, the warm-blooded mammals survived while the dinosaurs did not.

The carnivorous dinosaurs were cunning and aggressive, attacking prey with unusual voracity. The cranial capacity of some carnivores suggests they

Figure 136 *Many small dinosaurs such as* Mononykus *were built much like birds.*

Figure 137
Stenonychosaurus *had a
large brain weight to body
weight ratio.*

had relatively large brains and were fairly intelligent (Fig. 137), able to react to a variety of environmental pressures. The velociraptors, meaning "speedy hunters," had sharp claws and powerful jaws. They were vicious killing machines whose voracious appetites suggest they were warm-blooded.

Many dinosaur species were swift and agile, requiring a high rate of metabolism that only a warm-blooded body could provide. The anatomy of some dinosaurs implies they were active, upright, and constantly feeding, a behavior that suggests a high metabolic rate commensurate with warm-blooded animals. The comparison of ratios of oxygen 18 to oxygen 16 in dinosaur teeth with those of cold-blooded crocodiles suggests dinosaurs had a constant body temperature regardless of the outside temperature. They breathed with diaphragm-driven lungs for the extra respiration needed for rapid movement much like modern mammals. However, when at rest, they might have switched to rib-based respiration similar to reptiles, giving them a metabolism functioning unlike that of any living animal.

A comparison of the density of pelvic bones in fossilized dinosaur embryos with the bones of modern crocodiles and birds indicated that dinosaur juveniles were highly mobile, requiring the energy only a warm-blooded body could supply. Evidence for rapid juvenile growth, which is common among mammals, also exists in the bones of some dinosaur species, possibly providing another sign of warm-bloodedness. Newborn dinosaurs apparently were not defenseless dependents as with most birds but leapt from their shells ready to run and protect themselves from a dangerous world. The infant dinosaurs had the bone and muscle strength to move rapidly and scamper away from danger the way modern crocodile young do. However, a fossil discovery of *Maiasaura,* which literally means "good mother lizard," suggests that dinosaur adults were tender, nurturing parents whose young remained in their nest for feeding and protection.

The blood vessel density of dinosaur bones was higher than living mammals, another warm-blooded hallmark. Some dinosaur skulls display signs of sinus membranes, which exist only in warm-blooded animals. Respiratory turbinates, which act as thermal exchangers common in the nostrils of warm-blooded species, appear to have been present in dinosaurs as well. Bones of carnivorous dinosaurs found in Antarctica suggests that they either adapted to the cold and dark by being warm-blooded or migrated to nearby South America across a land bridge.

Dinosaur tracks (Fig. 138) are the most impressive of all fossil footprints because the great weight of many species produced deep indentations in the ground. Their footprints exist in relative abundance in terrestrial sediments of Mesozoic age throughout the world. The study of dinosaur tracks suggests that some species were highly gregarious and gathered in herds. Large carnivores such as *Tyrannosaurus rex* (Fig. 139), whose name means "terrible lizard," were swift, agile predators that could sprint as fast as a horse according to their tracks.

Females of some dinosaur species might have had live births like mammals. Many nurtured and fiercely protected their offspring until they could fend for themselves, allowing larger numbers to mature into adulthood, thus

Figure 138 Dinosaur trackways in the Chacarilla Formation, Tarapaca Province, Chile.

(Photo by R. J. Dingman, courtesy USGS)

Figure 139
Tyrannosaurus rex *was the greatest land carnivore that ever lived.*

ensuring the continuation of the species. The parents might have brought food to their young and regurgitated seeds and berries as do modern birds. This parental care for infants suggests strong social bonds and might explain why dinosaurs were so successful for so long.

Some dinosaurs might have developed complex mating rituals. Besides regulating its body temperature, the large sail on dimetrodon's back might have been used to attract females. Other dinosaurs might have sported elaborate headgear like present-day birds for much the same purpose. Dinosaurs might have used complex vocal resonating devices for mating calls. They appeared to produce sounds by oscillating cartilage in their throats. The larger the cartilage the lower the sound frequency, enabling some species to make a noise somewhat like a foghorn, which carried for a long distance. One 75-million-year-old herbivore had a 4.5-foot-long hollow crest arching backward from the skull. It was possibly used as a sexual display device or as a resonating chamber to produce unusual sound effects.

The period between the end of the Triassic and the beginning of the Jurassic was one of the most exciting times in the history of land vertebrates. In the late Triassic, all the continents were joined only at their western ends with Laurasia in the north and Gondwana in the south. Nonetheless, animal life on Laurasia was becoming distinct from that on Gondwana. A land bridge connecting Laurasia to the Indochina microcontinent might have been the last link, enabling the migration of animals when the two blocks collided as the Triassic drew to a close.

At the end of the Triassic, about 210 million years ago, a huge meteorite slammed into Earth, creating the 60-mile-wide Manicouagan impact structure in Quebec, Canada (Fig. 140). The gigantic explosion appears to have coincided with a mass extinction over a period of less than a million years that killed off 20 percent or more of all families of animals, including nearly half the reptile families. In the ocean, ammonoids and bivalves were decimated, and the conodonts completely disappeared. The extinction forever changed the character of life on Earth and paved the way for the rise of the dinosaurs.

Almost all modern animal groups, including amphibians, reptiles, and mammals, made their debut on the evolutionary stage at this time. This was also when the dinosaurs achieved dominance over Earth and held their ground for the next 145 million years. Afterward, another large meteorite

Figure 140 *The Manicouagan impact structure, Quebec, Canada.*

(Photo courtesy NASA)

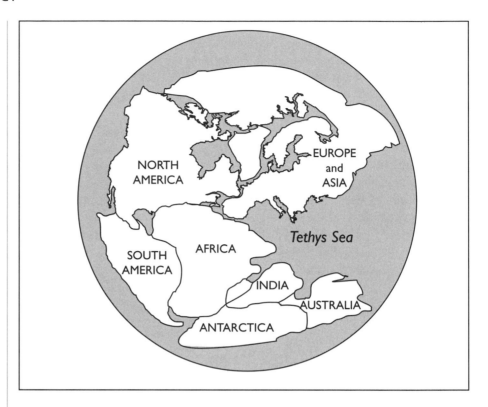

apparently struck Earth with the explosive force of 100 trillion tons of dynamite, turning the planet into an inhospitable world. Thus, the dinosaurs might have been both created and destroyed by meteorites.

THE TETHYAN FAUNA

At the beginning of the Triassic, the Tethys Sea was a huge embayment separating the northern and southern arms of Pangaea, which took the shape of a gigantic letter C that straddled the equator (Fig. 141). Between the late Paleozoic and middle Cenozoic, the Tethys was a broad tropical seaway that extended from western Europe to southeast Asia and harbored diverse and abundant shallow-water marine life. A great circumglobal ocean current that distributed heat to all parts of the world maintained warm climatic conditions. The energetic climate eroded many high mountain ranges of North America and Europe down to the level of the prevailing plain.

Early in the Triassic, ocean temperatures remained cool following the late Permian ice age. Marine invertebrates that managed to escape extinc-

190

tion lived in a narrow margin near the equator. Corals declined in the late Paleozoic and were replaced by sponges and algae when the seas they inhabited receded. This was due to extensive glaciation, whose ice caps captured a significant amount of Earth's water. Corals, which require warm, shallow water for survival, were particularly hard hit as evidenced by the lack of coral reefs at the beginning of the Triassic. When the great glaciers melted and the seas began to warm, reef building became intense in the Tethys Sea. Thick deposits of limestone and dolomite were laid down by prolific lime-secreting organisms.

The mollusks appeared to have weathered the hard times of the late Permian extinction quite well. They continued to become the most important shelled invertebrates of the Mesozoic seas, with some 70,000 distinct species living today. The warm climate of the Mesozoic influenced the growth of giant animals in the ocean as well as on land. Huge clams grew to 3 feet wide, giant squids were upward of 65 feet long and weighed over a ton, and tall crinoids reached 60 feet in length.

The cephalopods were extremely spectacular and diversified. They became the most successful marine invertebrates of the Mesozoic seas, evolving into some 10,000 species. The ammonoids evolved in the early Devonian 395 million years ago. The two forms best preserved in the fossil record were those with coiled shells up to 7 feet wide and more awkward types with straight shells growing to 12 feet in length. They traveled by jet propulsion, using neutral buoyancy to maintain depth, which contributed to their great success. However, of the 25 families of widely ranging ammonoids in the late Triassic, all but one or two became extinct at the end of the period. Those species that escaped extinction eventually evolved into the scores of ammonite families in the Jurassic and Cretaceous.

Among the marine vertebrates, fish progressed into more modern forms. The sharks regained ground lost from the great Permian extinction. They continued to become the successful predators in the oceans of today. Nevertheless, because of intense overfishing, some shark species appear to be headed for extinction.

During the final stages of the Cretaceous, when the seas departed from the land as the level of the ocean dropped, the temperatures in the Tethys Sea began to fall. As the continents drifted poleward during the last 100 million years, the land accumulated snow and ice, which had an additional cooling effect on the climate. Most warmth-loving species, especially those living in the Tethys Sea, disappeared. The most temperature-sensitive Tethyan fauna suffered the heaviest extinction rates. Species that were so successful in the warm waters of the Tethys dramatically declined when ocean temperatures dropped. Afterward, marine species acquired a more modern appearance as ocean bottom temperatures continued to plummet.

THE NEW RED SANDSTONE

The Triassic witnessed the complete retreat of marine waters from the land as the continents continued to rise. Abundant terrestrial redbeds and thick beds of gypsum and salt were deposited in the abandoned basins. Also, the amount of land covered with deserts was much greater in the Triassic than today. This is indicated by a preponderance of red rocks composed of terrestrial sandstones and shales exposed in the mountains and canyons in the western United States. Terrestrial redbeds covered a region from Nova Scotia to South Carolina and the Colorado Plateau (Fig. 142).

Redbeds are also common in Europe, where in northwestern England they are particularly well developed. Over northern and western Europe, the terrestrial redbeds are characterized by a nearly fossil-free sequence called the New Red Sandstone, named for a sedimentary formation in Scotland famous for its dinosaur footprints. The unit shows a continuous gradation from Permian to Triassic in the region, with no clear demarcation between the two periods.

The wide occurrences of red sediments probably resulted from massive accumulations of iron supplied by one the most intense intervals of igneous activity the world has ever known. Air trapped in ancient tree sap suggests a greater abundance of atmospheric oxygen, which oxidized the iron to form the mineral hematite, named so because of its blood red color.

Figure 142
Downwarping of the Pierce Canyon redbeds along the north margin of Nash Draw, Eddy County, New Mexico.

(Photo by J. D. Vine, courtesy USGS)

The mountain belts of the Cordilleran of North America, the Andean of South America, and the Tethyan of Africa-Eurasia contained thick marine deposits of Triassic age. The Cordilleran and Andean belts were created by the collision of east-Pacific plates with the continental margins of the American plates formed when Pangaea rifted apart. The Tethyan belt formed when Africa collided with Eurasia, raising the Alps, which contain an abundant fossil-bearing Triassic section. The Tethys Sea, located in the tropics during the Triassic, contained widespread coral reefs that uplifted to form the Dolomites and Alps during the collision of Africa and Eurasia in the Cenozoic.

During the Triassic, when North America was still part of Pangaea and the ocean began in present-day Nevada, an ancient major river system called the Chinle flowed through western United States. The source of the Chinle was the highlands of Texas, known as the Amarillo-Wichita Uplift. From there it flowed westward across northern New Mexico and southern Utah. It emptied into the sea at central Nevada. Sedimentary rocks of the Chinle Formation produced important uranium ore deposits responsible for the uranium boom of the American West beginning in the 1950s. Eventually, the Texas highlands eroded down to plains, and Pangaea drifted away from the equator into the drier midlatitudes. By the time North America separated from the other continents around 200 million years ago, the great Chinle River was drying up.

Late in the Triassic, an inland sea began to flow into the west-central portions of North America. Accumulations of marine sediments eroded from the Cordilleran highlands to the west, often referred to as the ancestral Rockies. They were deposited onto the terrestrial redbeds of the Colorado Plateau. Important reserves of phosphate used for fertilizers were precipitated in the late Permian and early Triassic in Idaho and adjacent states. Huge sedimentary deposits of iron were also laid down. The ore-bearing rocks of the Clinton Iron Formation, the chief iron producer in the Appalachian region from Alabama to New York, were deposited during this time.

Evaporite accumulation peaked during the Triassic, when the supercontinent Pangaea was just beginning to rift apart. Evaporite deposits form when shallow brine pools replenished by seawater evaporate. Few evaporite deposits date beyond 800 million years ago, however, probably because most of the salt was buried or recycled into the sea. Ancient evaporite deposits exist as far north as the Arctic regions. These indicate that either these areas were once closer to the equator or the global climate was considerably warmer in the past.

TRIASSIC BASALTS

Over the last 250 million years, 11 episodes of massive flood basalt volcanism have occurred worldwide (Fig. 143 and Table 9). They were relatively short-lived

Figure 143 *Many occurrences of flood basalt volcanism are associated with continental breakup.*

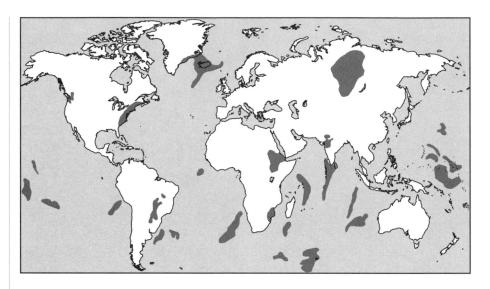

events, with major phases generally lasting less than 3 million years. These large eruptions created a series of overlapping lava flows, giving many exposures a terracelike appearance known as traps, from the Dutch word for "stairs."

In a geologically brief catastrophe at the end of the Triassic, massive rivers of basalt lava oozed out of giant fissures and paved over a continent-sized portion of land. The volcanic crisis occurred when all the continents

TABLE 9   FLOOD BASALT VOLCANISM AND MASS EXTINCTIONS

Volcanic Episode	Million Years Ago	Extinction Event	Million Years Ago
Columbia River, USA	17	Low-mid Miocene	14
Ethiopian	35	Upper Eocene	36
Deccan, India	65	Maastrichtian	65
		Cenomanian	91
Rajmahal, India	110	Aptian	110
South-West African	135	Tithonian	137
Antarctica	170	Bajocian	173
South African	190	Pliensbachian	191
E. North American	200	Rhaectian/Norian	211
Siberian	250	Guadalupian	249

huddled together into a single landmass named Pangaea. Within a short span of a few million years, black basalt erupted along a central spine of the super-continent, eventually spreading over an area nearly the size of Australia. Soon thereafter, the continents split apart along this axis and went their separate ways, thereby opening up the Atlantic Ocean. This massive eruption, one of the largest in Earth history, might have killed off much of the planet's life and led to the ascension of the dinosaurs as the rulers of the world.

Many flood basalts lie near continental margins, where great rifts began to separate the present continents from Pangaea near the end of the Triassic. These massive outpourings of basalt reflected one of the greatest crustal movements in the history of the planet. The continents probably traveled much faster than they do today because of more vigorous plate motions, resulting in tremendous volcanic activity.

Triassic basalts common in eastern North America indicate the formation of a rift that separated the continent from Eurasia. The rift later breached and flooded with seawater, forming the infant North Atlantic Ocean. The Indian Ocean formed when a rift separated the Indian subcontinent from Gondwana. By the end of the Triassic, India had drifted free and began its trek toward southern Asia. Meanwhile, Gondwana drifted northward, leaving Australia still attached to Antarctica behind in the South Polar region.

Huge basalt flows and granitic intrusions occurred in Siberia. Extensive lava flows covered South America, Africa, and Antarctica as well. Southern Brazil was paved with three-quarters of a million square miles of basalt, constituting one of the largest lava fields in the world. Great floods of basalt, upward of 2,000 or more feet thick, covered large parts of Brazil and Argentina when the South American plate overrode the Pacific plate, and the subduction fed magma chambers underlying active volcanoes. Basalt flows also blanketed a region from Alaska to California.

Near the end of the Triassic, North and South America began to move away from each other. India, nestled between Africa and Antarctica, began to separate from Gondwana. The Indochina block collided with China. Additionally, a great rift began to divide North America from Eurasia. The rifting of continents radically altered the climate and set the stage for the extraordinary warm periods that followed.

After examining the early dinosaurs of the Triassic, the next chapter explores the animals that took to the air during the Jurassic period.

11

JURASSIC BIRDS
THE AGE OF FLYING CREATURES

This chapter examines the creatures that took flight and covers the largest animals ever to inhabit Earth during the Jurassic period. The Jurassic, from 210 to 135 million years ago, was named for the limestones and chalks of the Jura Mountains in northwest Switzerland. Early in the period, Pangaea began rifting apart into the present-day continents, forming the Atlantic, Indian, and Arctic Oceans. Mountains created by crustal upheavals during previous periods were leveled by erosion. Inland seas invaded the continents, providing additional offshore habitats for a bewildering assortment of marine species (Fig. 144). By this time, terrestrial faunas had attained the basic composition they would keep until the dinosaurs became extinct.

The dinosaurs were highly diversified during the Jurassic and reached their maximum size, becoming the largest terrestrial animals ever to live. Widespread plant growth and coal formation suggest a warm, moist climate. The beneficial climate and magnificent growing conditions contributed to the giant size of some dinosaur species, many of which became extinct at the end of the period. Reptiles were extremely successful and occupied land, sea, and air. Mammals were small, rodentlike creatures, sparsely populated and

Figure 144 *Marine flora and fauna of the late Jurassic.*

(Courtesy Field Museum of Natural History)

scarcely noticed. The first flight-worthy birds appeared and shared the skies with flying reptiles called pterosaurs.

THE EARLY BIRDS

Birds first appeared in the Jurassic about 150 million years ago, although some accounts push their origin as far back as the late Triassic 225 million years ago. By about 135 million years ago, early birds began to diversify. They diverged into two lineages. One led to archaic birds, and the other led to modern birds. The birds are believed to have descended from the diapsid reptiles called thecodonts, the same ancestors of dinosaurs and crocodiles. Consequently, birds are often referred to as "glorified reptiles." Alternatively, they could have

Figure 145 *An oviraptor brooding a clutch of eggs.*

instead descended from bipedaled, meat-eating dinosaurs known as theropods, from Greek meaning "fierce foot."

Indeed, birds are the only living relatives of dinosaurs. Except for having a long tail, the skeletons of many small dinosaur species closely resembled those of birds, suggesting a direct descent from dinosaurs. Microraptor is by far the smallest adult dinosaur yet discovered and was about the size of a crow. The creature had feathers on its body. It also had skeletal features that linked it with birds, including birdlike clawed feet that might have been used for perching on branches. The animal is considered the closest dinosaurian relative of birds and was probably a nimble climber of trees. Its ancestors might have lived in trees long enough to develop traits associated with an arboreal lifestyle. Therefore, flying would have been a simple matter of launching from the treetop back to the ground.

Oviraptors (Fig. 145) are dinosaurs that brooded their eggs in a communal nest. They are closely related to birds, which provide the strongest evidence of parental attention, suggesting birds inherited this behavior from dinosaurs. Oviraptors and birds probably evolved from a common ancestor that also exhibited brooding behavior. The animal apparently lived prior to the earliest birds, which evolved alongside the dinosaurs, who mysteriously died out while birds live on.

Birds are warm-blooded to obtain the maximum metabolic efficiency needed for sustained flight but have retained the reptilian mode of reproduction by laying eggs. The bones of some Cretaceous birds show growth rings, a feature common among cold-blooded reptiles. This suggests that early birds might not have yet developed fully warm-blooded bodies. Birds' ability to maintain high body temperatures has sparked a controversy over whether some dinosaur species with similar skeletons were warm-blooded as well.

Archaeopteryx (Fig. 146), from Greek meaning "ancient wing," was the earliest known fossil bird. It was about the size of a modern pigeon and appeared to be a species in transition between reptiles and true birds. Unlike modern birds, it lacked a keeled sternum for the attachment of flight muscles. *Archaeopteryx* was first thought to be a small dinosaur until fossils clearly showing impressions of feathers were found in a unique limestone formation in Bavaria, Germany, in 1863. The discovery sparked a long-standing controversy. Prominent 19th-century geologists claimed *Archaeopteryx* was a hoax and that the feather impressions were simply etched into the rock containing the fossil. However, an *Archaeopteryx* fossil discovered in 1950 from the same Bavarian formation produced a well-preserved specimen that clearly showed feather impressions.

Although *Archaeopteryx* had many accoutrements necessary for flight, it was likely a poor flyer. It might have flown only short distances, similar to today's domesticated birds. It probably achieved flight by running along the ground with its wings outstretched and glided for a brief moment or leaped from the ground while flapping its wings to catch an insect flying by. Their

Figure 146
Archaeopteryx *is a possible link between reptiles and birds.*

forbearers might have flapped their wings to increase running speed while escaping predators, thereby obtaining flight purely by accident.

Archaeopteryx had teeth, claws, a long bony tail, and many skeletal features of a small dinosaur but lacked hollow bones for light weight. Its feathers were outgrowths of scales and probably originally functioned as insulation. Some bird species retained their teeth until the late Cretaceous roughly 70 million years ago. Claws on the forward edges of the wings might have helped the bird climb trees, from which it could launch itself into the air.

Upon mastering the skill of flight, birds quickly radiated into all environments. Their superior adaptability enabled them to compete successfully with the pterosaurs, possibly leading to that reptile's extinction. Giant flightless land birds appeared early in the avian fossil record. Their wide distribution is further evidence for the existence of Pangaea since these birds would then have had to walk from one corner of the world to another.

After being driven into the air by the dinosaurs and kept there by hunting mammals, birds found life much easier on the ground once this threat was eliminated because they had to expend a great deal of energy to remain airborne. Some birds also successfully adapted to a life in the sea. Certain diving ducks are specially equipped for "flying" underwater to catch fish. Penguins, for example, are flightless birds that have taken to life in the water and are well adapted to survive in the cold Antarctic.

THE PTEROSAURS

During the Jurassic, reptiles achieved great success and dominated land, sea, and air. Some 160 million years ago, when the dinosaurs were the dominant species on the continents, remarkable fish-shaped reptiles called ichthyosaurs reigned over the oceans. They belonged to a group of animals known as diapsids, which includes snakes, lizards, crocodiles, and dinosaurs. Ichthyosaurs ruled the seas from the time the dinosaurs first appeared about 245 million years ago to about 90 million years ago. The largest of these animals were whale sized, exceeding 50 feet in length. They had enormous eyes as big around as bowling balls. Their fossils have been found throughout the world, suggesting they migrated extensively like modern whales.

Overhead were flying reptiles with wingspans up to 30 feet or more called pterosaurs (Fig. 147), including the ferocious-looking pterodactyl. They originated in the early Jurassic and appear to be the largest animals that ever flew, dominating the skies for more than 120 million years. Pterosaurs resembled both birds and bats in their overall structure and proportions, with the smallest species roughly the size of a sparrow. As with birds, they had hollow bones to conserve weight for flight. The larger pterosaurs were proportioned

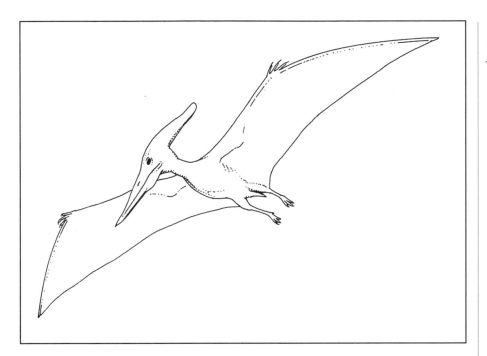

Figure 147 *The giant pterosaurs ruled the skies for 120 million years.*

similar to a modern hang glider and weighed about as much as the human pilot. Many pterosaurs had tall crests on their skulls, which possibly functioned as a forward rudder to steer it in flight.

The pterosaur wings were constructed by elongating the fourth finger of each forelimb, which supported the front edge of a membrane that stretched from the flank of the body to the fingertip. The other fingers were left free for such purposes as climbing trees. By comparison, a bat's wing is constructed by lengthening and splaying all fingers and covering them with membrane.

The first reconstructions of a pterodactyl portrayed the animal as an ungainly glider with leathery wings attached on each side of the body down to the legs and stretched between a grotesquely elongated finger. Such a large flap of skin would have made the creature a clumsy walker, shuffling around on all fours like a bat, not a good way to take off. Later reconstructions showed a flying reptile with thin birdlike wings attached to the hip, allowing the animal to walk on two legs, a better design for gaining speed for liftoff.

Why pterosaurs took to the air in the first place remains a mystery. They apparently arose from tree-dwelling reptiles rather than from running ground dwellers. Their ancestors might have grown skin flaps for jumping from tree to tree in a similar manner as flying squirrels. The wing membranes might have originally served as a cooling mechanism that regulated body temperature by fanning the forelimbs, and through natural selection they eventually became flying appendages.

The pterosaur might have achieved flight by jumping off cliffs and riding the updrafts, by climbing trees and diving into the wind, or by gliding across the tops of wave crests like modern albatrosses. The animal could have trotted along the ground flapping its wings and taking off gooney bird fashion. It could have simply stood on its hind legs, caught a strong breeze, and with a single flap of its huge wings and a kick of its powerful legs became airborne. It probably spent most of its time aloft riding air currents like present-day condors.

When landing, it simply stalled near the ground, gently touching down on its hind legs like a hang glider does. While on the ground, pterosaurs might have been ungainly walkers, sprawling about on all fours similar to the way a bat does. However, fossil pterosaur pelvises seem to indicate that the hind legs extended straight down from the body, enabling the reptiles to walk upright on two feet. They could then trot along for short bursts to gather speed for takeoff.

Pterosaurs living in North America about 75 million years ago were thought to be toothless, until the first evidence of a flying reptile with teeth approximately an inch long was discovered in Texas. The fossil record of pterosaurs is sketchy because their delicate bones were easily destroyed. The only remains found of the Texas specimen was its bill, which had a distinctive crest that apparently helped stabilize the pterosaur's head like a keel as it dipped into the sea to catch fish.

The discovery of the 100-million-year-old Texas pterosaur proves that teeth-bearing flying reptiles similar to those in South America and Eurasia migrated to North America. The pterosaur had a wingspan of about 5 feet and a slender body about 1.5 feet long. It resembled a species that lived in Great Britain about 140 million years ago. Whether any connection exists between this toothy North American pterosaur and the toothless variety that came later is still unclear.

The largest pterosaur was about the size of a small airplane, with a wingspan of 30 feet. Fossils of the huge flier were unearthed in Big Ben National Park in southwest Texas. Originally thought to feed on carrion such as dead dinosaurs, it was most likely a fish eater resembling a giant pelican even though the deposits where the animal was found were far from the sea. Their extremely large size enabled them to fly long distances to the ocean, where they could find plentiful fish to catch. That the animals could actually fly leaves little doubt, and they went on to become the world's greatest aviators.

THE GIANT DINOSAURS

Dinosaurs attained their largest sizes and longest life spans during the Jurassic. The biggest dinosaurs occupied Gondwana, which included all the southern continents. The Jurassic Morrison Formation, a famous bed of sediments in

Figure 148 *Restoration of dinosaur bones at Dinosaur National Monument, Utah.*

(Courtesy National Park Service)

the Colorado Plateau region, has yielded some of the largest dinosaur fossils. Many of the best specimens are displayed at Dinosaur National Monument near Vernal, Utah (Fig. 148).

More than 500 dinosaur genera have been discovered over the last two centuries. Among the largest dinosaur species were the sauropods. These gargantuan creatures fully deserve the title "thunder lizard." They had long, slender tails and necks, and the front legs were generally longer than the hind legs. Their fossils are found in Colorado and Utah, southwestern Europe, and East Africa, with some species probably having traveled to Africa by way of Europe.

Dinosaurs are often portrayed as lumbering, unintelligent brutes. Yet the fossil record suggests that many species, such as velociraptor, were swift and intelligent. All dinosaurs were not giants, and many species were no larger than most mammals today. The *Protoceratops*—a parrot-beaked, shield-headed dinosaur—and the ankylosaur, whose name means "stiff lizard"—an elephant-sized dinosaur heavily covered with protective bone and swinging a club-shaped tail—were very common and ranged throughout the world.

Some large bipedaled dinosaurs later reverted to a four-footed stance as their weight increased. They eventually evolved into gigantic, long-tailed, long-necked sauropods such as the apathosaurs (Fig. 149), which belong to the

Figure 149 Dinosaurs such as apathosaur reached their maximum size during the Jurassic period.

same family of dinosaurs as the brontosaurs. Others, such as *Tyrannosaurus rex,* perhaps the fiercest land carnivore of all time, maintained a permanent two-legged stance with powerful hind legs, a muscular tail for counterbalance, and arms shortened to almost useless appendages.

Tyrannosaurus rex was at a distinct disadvantage as a galloping predator, however. It was bipedaled with very strong legs, but its weak forelimbs were unable to break a fall. If a 7-ton *T. rex* charging prey at a top speed of 25 miles per hour were to trip and fall, the crash could be fatal. More likely, the dinosaurs traveled in hunting packs that surrounded their prey. This would have required much less speed, thereby resulting in fewer crashes. About 90 million years ago, when the continents became isolated and evolved separate faunas, *Tyrannosaurus rex* began to roam the American West.

Oviraptor was a fleet-footed predator, whose name, which literally means "egg hunter," was a misnomer because it was originally thought to have raided nests of other dinosaurs. A fossil oviraptor was found sitting on a nest filled with as many as two dozen eggs neatly laid out in a circle, with the thinner ends pointing to the outside. The dinosaur resembled a wingless version of an ostrich with a shortened neck and a long tail. It sat with its pelvis in the middle of the nest and had its long arms wrapped around the nest the way birds do.

The oviraptor was perhaps protecting the eggs against a gigantic sand-storm that apparently engulfed and fossilized it along with its clutch 70 to 80 million years ago. The oviraptor was in the exact position a chicken would take sitting on a nest. Whether the dinosaur was keeping the eggs warm as birds do, shading them from the hot sun, or protecting them is unknown. The clutch was probably a communal nest similar to that used by ostriches, into which hens deposit their eggs and take turns incubating them.

The giant herbivores might have traveled in great herds, with the largest adults in the lead and the juveniles placed in the center for protection. Duck-billed hadrasaurs (Fig. 150), among the most successful of all dinosaur groups, were up to 15 feet tall and lived in the Arctic regions. There they had to adapt to the cold and dark or else migrate in large herds over long distances to warmer climates. Triceratops, whose vast herds roamed the entire globe toward the end of the Cretaceous and were among the last to go during the dinosaur extinction, might have contributed to the decline of other species of dinosaurs possibly due to extensive habitat destruction or the spread of diseases.

Fossils of western North American triceratops have been found in sediments deposited between 73 and 65 million years ago, when the region was covered by a warm, shallow sea. Parallel sets of footprints and bone beds with large numbers of remains suggest that triceratops congregated in large herds that might have migrated over long distances. Rather than walk-ing with a sprawling, crocodilelike gait as originally thought, the animal walked erect similar to a rhino. It could probably run as fast as one. It could also lock its knees like cows and horses do when they sleep standing up. The dinosaur was one of the dominant herbivores long before grasses evolved. It had an enormous head, weighing nearly half a ton and used its turtlelike beak and bladelike teeth to graze on flowering shrubs, palm fronds, and small trees.

Figure 150 Hadrasaurs lived in the Arctic regions during the Cretaceous.

Figure 151
Giganotosaurus *was the largest carnivorous dinosaur.*

Among the largest dinosaur species were the apathosaurs and brachiosaurs, which lived around 100 million years ago. Paralit%titan, meaning "tidal giant," was the second largest dinosaur, measuring 100 feet in length and weighing 75 tons. It resembled apathosaurs and lived in a lush coastal region 94 million years ago. Perhaps the tallest and heaviest dinosaur thus far discovered was the 80-ton *Ultrasaurus,* which could tower above a five-story building. *Seismosaurus,* meaning "earth shaker," was the longest known dinosaur. It reached a length of more than 140 feet from its head, which was supported by a long, slender neck, to the tip of its even longer whiplike tail. *Giganotosaurus* (Fig. 151) even challenges *Tyrannosaurus rex* as the most ferocious terrestrial carnivore that ever lived.

Large reptiles possess the power of almost unlimited growth. Adults never cease growing entirely but continued to increase in size until disease, accident, or predation takes their lives. The giant Komodo dragon lizards of southeast Asia, for example, grow to over 300 pounds and prey on monkeys, pigs, and deer. Reptiles with their continuous growth achieve a measure of eternal youth, whereas mammals grow rapidly to adulthood and then slowly degenerate and die.

A large body allows a cold-blooded reptile to retain its body temperature for long periods. A large body retards heat loss better than a small one because it has a better surface area to volume ratio. Thus, the animal is less susceptible to short-term temperature variations such as cool nights or cloudy days. Conversely, a large reptile takes much longer to warm up from an extended cold period than a small one. Muscles also generate body heat, although for reptiles they only produce about a quarter of that of mammals during exertion. High steady body temperatures maintain an efficient metabolism and enhance the output of muscles. Therefore, the performance of some large dinosaurs probably could match that of large mammals.

The generally warm climate of the Mesozoic produced excellent growing conditions for lush vegetation, including ferns and cycads, to satisfy the diets of the plant-eating dinosaurs. Much information can be obtained about dinosaur diets by studying coprolites (Fig. 152), which are masses of fecal matter preserved as fossils and are usually modular, tubular, or pellet shaped. Dinosaur coprolites can be quite massive, even larger than a loaf of bread.

Coprolites are often used to determine the feeding habits of extinct animals. For example, coprolites of herbivorous dinosaurs are black and blocky shaped and usually filled with plant material. Those of carnivorous dinosaurs are spindle shaped and contain broken bits of bone from dining on other animals. Some dinosaur species swallowed cobbles called gizzard stones, similar to the grit used by modern birds in order to grind the vegetation in their stomachs into pulp. The rounded, polished stones called gastroliths were left in a heap where the dinosaur died. Sometimes deposits of these stones are found atop exposed Mesozoic sediments.

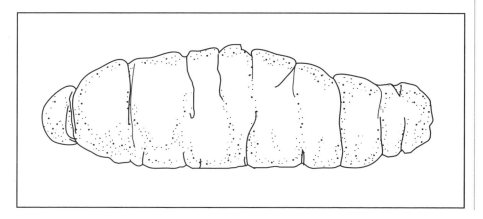

Figure 152 Coprolites are fossilized dinosaur droppings.

The giant herbivorous dinosaurs such as apathosaurs and stegosaurs (Fig. 153) developed a large stomach to digest the tough, fibrous fronds, requiring an enormous body to carry it around. The dinosaurs grew to such giants probably for the same reasons that large ungulates such as the rhinoceros and elephant are so big. Most large dinosaurs were herbivores that consumed huge quantities of coarse cellulose that required much time to digest. The digestive juices further broke down the rough material, and the long fermentation process required a large storage capacity.

The large size of the herbivores spurred the evolution of giant carnivorous dinosaurs to prey on them such as *Tyrannosaurus rex,* perhaps the fiercest carnivore of them all. The giant dinosaurs were prevented from growing any larger due to the force of gravity. When an animal doubles its size, the weight on its bones is four times greater. The only exceptions were dinosaurs living permanently in the sea. As with modern whales, some of which are even larger than the biggest dinosaurs, the buoyancy of seawater kept the weight off their bones. If the animal accidentally beached itself, as whales sometimes do, it suffocated because its bones were unable to support the weight of its body, crushing its lungs.

Figure 153 *Stegosaurs were ornithischian dinosaurs with strongly developed dorsal plates and spikes and known from upper Jurassic rocks of Colorado and Wyoming.*

Figure 154 *Allosaurs were among several dinosaur genera that went extinct at the end of the Jurassic.*

(Photo courtesy National Museums of Canada)

Many families of large dinosaurs, including apathosaurs, stegosaurs, and allosaurs (Fig. 154) became extinct at the end of the Jurassic. The allosaur was a carnivorous dinosaur with a thick skull to withstand a high-impact attack on prey. The dinosaur would ambush its victim, slam into it at high speed with jaws wide open, and drive its sharp teeth into the flesh.

Following the large dinosaur extinction, the population of small animals exploded, as species occupied niches vacated by the large dinosaurs. Most of the surviving species were aquatic, confined to freshwater lakes and marshes, and also small, land-dwelling animals. Many of the small, nondinosaur species were the same types that survived the dinosaur extinction at the end of the Cretaceous, probably due to their large populations and ability to find refuge during climatic disturbances.

THE BREAKUP OF PANGAEA

Throughout Earth's history, continents appear to have undergone cycles of collision and rifting. Smaller continental blocks collided and merged into larger continents. Millions of years later, the continents rifted apart, and the chasms filled with seawater to form new oceans. The regions presently bordering the Pacific Basin apparently have not collided with each other. The Pacific Ocean is a remnant of an ancient sea called the Panthalassa, which narrowed and widened in response to continental breakup, dispersal, and reconvergence in the area occupied by the present-day Atlantic Ocean.

Several oceans have repeatedly opened and closed in the vicinity of the Atlantic Basin, while a single ocean has existed continuously in the area of the Pacific Basin. The Pacific plate was hardly larger than the United States after the breakup of Pangaea in the early Jurassic about 180 million years ago. The rest of the ocean floor consisted of other unknown plates that disappeared as the Pacific plate grew. Consequently, no oceanic crust is older than Jurassic in age.

The Jurassic was a tumultuous time. Pangaea split apart into the present-day continents, leaving a gaping rift that filled with seawater to become the Atlantic Ocean. Sea level and climate changed dramatically as the continents tore away from each other. Yet life endured this tremendous geologic upheaval for millions of years with little effect.

When Pangaea began to separate into today's continents (Fig. 155), a great rift developed in the present Caribbean. It sliced northward through the continental crust connecting North America, northwest Africa, and Eurasia and began to open the Atlantic Ocean. The process took several million years along a zone hundreds of miles wide. The breakup of North America and Eurasia might have resulted from upwelling basaltic magma that weakened the continental crust. Many flood basalts exist near continental margins, where rifts separated the present continents. The episodes of flood basalt volcanism were relatively short-lived events, with major phases generally lasting less than 3 million years.

India, nestled between Africa and Antarctica, drifted away from Gondwana. Antarctica, still attached to Australia, swung away from Africa toward the southeast, forming the proto–Indian Ocean. The rift separating the continents breached and flooded with seawater, forming the infant North Atlantic Ocean. Many ridges of the Atlantic's spreading seafloor remained above sea level, creating a series of stepping-stones for the migration of animals between the Eastern and Western Hemispheres.

About 125 million years ago, the infant North Atlantic obtained a depth of about 2.5 miles and was bisected by an active midocean ridge system producing new oceanic crust. At about the same time, the South Atlantic began to form, opening like a zipper from south to north. The rift propagated northward several inches per year, comparable to the rate of separation between the plates. The entire process of opening the South Atlantic was completed in only about 5 million years. By 80 million years ago, the North Atlantic had become a fully developed ocean. Some 20 million years later, the Mid-Atlantic rift progressed into the Arctic Basin, detaching Greenland from Europe.

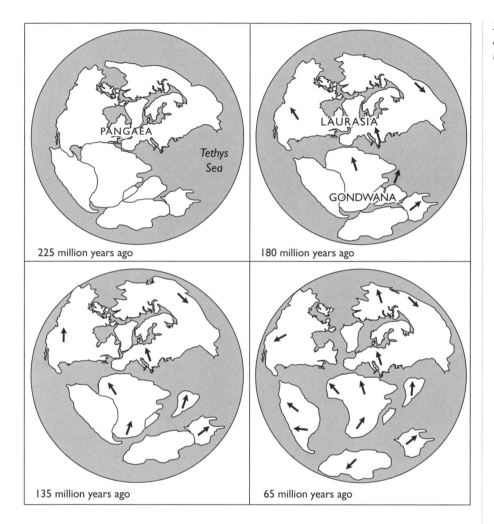

After the breakup, the continents traveled in spurts rather than drifting apart at a constant speed. The rate of seafloor spreading in the Atlantic was matched by plate subduction in the Pacific, where one plate dives under another, forming a deep trench. This is why the oceanic crust of the Pacific Basin dates back no farther than the early Jurassic. A high degree of geologic activity around the rim of the Pacific Basin produced practically all the mountain ranges facing the Pacific and the island arcs along its perimeter.

Much of western North America assembled from island arcs and other crustal debris skimmed off the Pacific plate as the North American plate continued heading westward. Northern California is a jumble of crustal fragments assembled within about 200 million years ago. A nearly complete slice of ocean crust, the type shoved up on the continents by drifting plates, sits in the

Figure 156 *A view across Kern Canyon, with Kaweah Peaks Ridge in the background in the Sierra Nevada, Sequoia National Park, Tulare County, California.*

(Photo by F. E. Matthes, courtesy USGS)

middle of Wyoming. The Nevadan orogeny produced the Sierra Nevada in California (Fig. 156) during the middle to late Jurassic.

The breakup of Pangaea compressed the ocean basins, causing a rise in sea level and a transgression of the seas onto the land. In addition, an increase in volcanism flooded the continental crust with vast amounts of basalt. The rise in volcanic activity also increased the carbon dioxide content of the atmosphere, resulting in a strong greenhouse effect that led to the warm Mesozoic climate.

Continental breakup and dispersal might have also contributed to the extinction of many dinosaur species. The shifting of continents changed global climate patterns and brought unstable weather conditions to many parts of the world. Massive lava flows from perhaps the most volcanically active period since the early days of Earth might have dealt a major blow to the climatic and ecological stability of the planet.

MARINE TRANSGRESSION

Throughout most of Earth's history, several crustal plates constantly in motion reshaped and rearranged continents and ocean basins. When continents broke up, they overrode ocean basins, which compressed the seas and made them less confined to their basins, thereby raising global sea levels several hundred feet. The rising sea inundated low-lying areas inland of the continents, dramatically increasing the shoreline and shallow-water marine habitat area, which in turn supported many more species.

The vast majority of marine species live on continental shelves, shallow-water portions of islands, and subsurface rises generally less than 600 feet deep. The richest shallow-water faunas live in the tropics, which contain large numbers of highly specialized organisms. Species diversity also depends on the shapes of the continents, the width of shallow continental margins, the extent of inland seas, and the presence of coastal mountains that supply nutrients to the ocean, all of which are affected by continental motions.

Extensive mountain building is also associated with the movement of crustal plates. The upward thrust of continental rocks alters patterns of river drainage and climate, which in turn affects terrestrial habitats. The raising of land to higher elevations, where the air is thin and cold, spurs the growth of glacial ice, especially in the higher latitudes. Furthermore, continents scattered in all parts of the world interfere with ocean currents, which distribute heat over the globe.

During the Jurassic and continuing into the Cretaceous, an interior sea flowed into the west-central portions of North America (Fig. 157). Massive

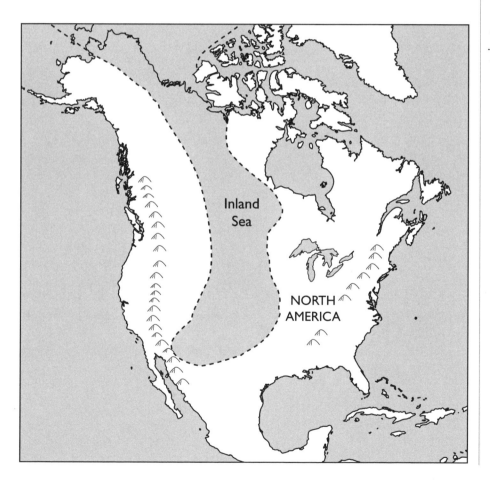

Figure 157 *The middle Jurassic inland sea in North America.*

Figure 158 *A dinosaur boneyard at the Howe Ranch quarry near Cloverly, Wyoming.*

(Photo by G. E. Lewis, courtesy USGS)

accumulations of marine sediments eroded from the Cordilleran highlands to the west were deposited on the terrestrial redbeds of the Colorado Plateau, forming the Jurassic Morrison Formation, well-known for fossil bones of large dinosaurs (Fig. 158). Eastern Mexico, southern Texas, and Louisiana were also flooded. Seas invaded South America, Africa, and Australia as well.

The continents were flatter, mountain ranges were lower, and sea levels were higher. Thick deposits of sediment that filled the seas flooding North America were uplifted and eroded, giving the western United States its impressive scenery. Reef building was intense in the Tethys Sea. Thick deposits of limestone and dolomite were laid down in the interior seas of Europe and Asia, later to be uplifted during one of geologic history's greatest mountain building episodes.

After covering the birds and giant dinosaurs of the Jurassic, the next chapter explores the life-forms and landforms of the warm Cretaceous period and the extinction of the dinosaurs.

12

CRETACEOUS CORALS
THE AGE OF TROPICAL BIOTA

This chapter examines the life-forms and landforms of the Cretaceous period and the extinction of the dinosaurs. The Cretaceous, from 135 to 65 million years ago, was named for the Latin word *creta,* meaning "chalk" due to vast deposits of carbonate rock laid down worldwide. It was the warmest period of the Phanerozoic as evidenced by extensive coral reefs, which built massive limestone deposits (Fig. 159). Coral and other tropical biota, for which bright sunlight and warm seas are essential, ranged far into high latitudes. The reefs fringed the continents and covered the tops of extinct marine volcanoes.

The warm climate was particularly advantageous to the ammonites, which grew to tremendous size. They became the predominant creatures of the Cretaceous seas. The dinosaurs did exceptionally well during the Cretaceous. However, along with the ammonites and many other species, they mysteriously vanished at the end of the period. The extinction was apparently caused by some sort of cataclysm that created intolerable living conditions for most species on Earth.

Figure 159 *Limestones of the Hawthorn and Ocala formations, Marion County, Florida.*

(Photo by G. H. Espenshade, courtesy USGS)

THE AMMONITE ERA

Coral reefs were the most widespread during the Cretaceous, ranging 1,000 miles away from the equator. In contrast, today they are restricted to the tropics. The corals began constructing reefs in the early Paleozoic and built barrier reefs and atolls, which were massive structures composed of calcium carbonate lithified into limestone. The Great Barrier Reef, stretching more than 1,200 miles along the northeast coast of Australia, is the largest feature built by living organisms.

Bryozoans, corallike moss animals that encrusted shells and other hard surfaces of the early Cretaceous seas, had been around for more than 300 million years. Like most shallow-water marine invertebrates, bryozoans were rare in the Triassic. However, they underwent widespread expansion in the Jurassic and Cretaceous until the late Cretaceous. Early in the Cretaceous, the bryozoans evolved into two major groups, the cyclostomes and the newer cheilostomes, whose higher growth rate and greater diversity crowded out the older group, which was relegated to inferior status. The typical encrusting forms had become a major group of bryozoans by the end of the Cretaceous, a position they maintain today. Living species occupy seas at various depths, with certain rare members adapted to life in freshwater. They have existed since the Cretaceous about 140 million years ago.

Sponges were common reef builders in the tropical Tethys Sea during the Mesozoic, particularly in the Cretaceous. Likewise, hexacorals, which ranged from the Triassic to the present and were the major reef builders of the Mesozoic and Cenozoic seas, experienced their greatest abundance and diversity in the Tethys. The brachiopods reached their peak in the Jurassic and declined thereafter. The gastropods, including snails and slugs, were most abundant in the Cretaceous, when the modern carnivorous types appeared. Gastropods increased in number and variety throughout the Cenozoic and are presently second only to insects in diversity.

The rudists were a significant group of reef-dwelling clams restricted to seas of the upper Jurassic and Cretaceous. The echinoids, including the urchins, became abundant for the first time in the Jurassic and Cretaceous but have declined to obscurity since then. The tubes of annelid worms are particularly common in Cretaceous marine strata. The star-shaped columnals of certain crinoids (Fig. 160) are occasionally common in Triassic and Jurassic rocks. The floating-swimming crinoids underwent a short, widespread evolutionary burst in the late Cretaceous, which makes them useful for dating rocks of the period.

Figure 160 Cogwheel-shaped crinoid columnals in a limestone bed of the Drowning Creek Formation, Flemming County, Kentucky.

(Photo by R. C. McDowell, courtesy USGS)

Figure 161 Cretaceous ammonite fossils on display at the Museum of Geology, South Dakota School of Mines at Rapid City.

The cephalopods were the most spectacular, diversified, and successful marine invertebrates of the Mesozoic seas. The nautiloids grew to lengths of 30 feet or more. Because of their straight, streamlined shells, they were among the swiftest creatures of the deep. The ammonites, the most significant cephalopods, had a variety of coiled-shell forms (Fig. 161) identified by their complex suture patterns. This makes them the most important guide fossils for dating Mesozoic rocks.

The ammonite shells were subdivided into air chambers. The suture lines joining the segments presented a variety of patterns used for identifying various species. The air chambers provided buoyancy to counterbalance the weight of the growing shell. Most shells were coiled in a vertical plane. Some forms were spirally coiled. Others were essentially straight, which often made swimming awkward. Their large variety of coiled-shell forms made the ammonoids ideal for dating Paleozoic and Mesozoic rocks. Throughout the Mesozoic, ammonite shell designs steadily improved. The cephalopod became one of the swiftest creatures of the deep sea, competing successfully with fish.

For 350 million years, 10,000 ammonite species roamed the seas. Of the 25 families of widely ranging ammonoids living in the late Triassic, however, all but one or two became extinct at the end of the period, when half of all species died out. The ammonites that managed to escape extinction eventually evolved into dozens of ammonite families during the Jurassic and Cretaceous. The ammonites lived mainly at the middle depths and might have shared many features with living squid and cuttlefish. The nautilus, commonly referred to as a living fossil because it is the ammonite's only living relative, lives at extreme depths of 2,000 feet.

The belemnoids, with long, bulletlike shells, originated from more primitive nautiloids and were related to the modern squid and octopuses. They were abundant during the Jurassic and Cretaceous and became extinct by the Tertiary. The shell was straight in most species and loosely coiled in others. The chambered part of the shell was smaller than that of the ammonoid, and the outer walls thickened into a fat cigar shape.

Unfortunately, after surviving the critical transition from the Permian to the Triassic and recovering from serious setbacks during the Mesozoic, the ammonites suffered final extinction at the end of the Cretaceous. At that time, the recession of the seas reduced their shallow-water habitats worldwide. The ammonites declined over a period of about 2 million years, possibly becoming extinct 100,000 years before the end of the Cretaceous.

A fast-swimming, shell-crushing marine predator called ichthyosaur apparently preyed on ammonites by first puncturing the shell from behind, causing it to fill with water and sink to the bottom. Then, the attack could be made on the more vulnerable front side of the ammonite. These highly aggres-

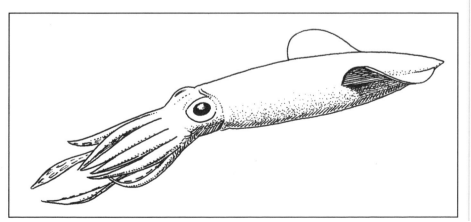

Figure 162 *Squid were among the most successful cephalopods.*

sive predators might have caused the extinction of most ammonite species before the Cretaceous was over.

All shelled cephalopods were absent in the Cenozoic seas except the nautilus. It is found exclusively in the deep waters of the Indian Ocean. The nautilus is the ammonite's only living relative, along with shell-less species, including cuttlefish, octopuses, and squids (Fig. 162). Squids competed directly with fish, which were little affected by the extinction. Other major marine groups that disappeared at the end of the Cretaceous include the rudists, which were huge, coral-shaped clams, and other types of clams and oysters.

THE ANGIOSPERMS

The Mesozoic was a time of transition, especially for plants. They showed little resemblance at the beginning of the era to those at the end, when they more closely resembled present-day vegetation. The gymnosperms, including conifers, ginkgoes, and palmlike cycads, originated in the Permian and bore seeds without fruit coverings. The true ferns prospered in the higher latitudes, whereas today they live only in the warm tropics.

The cycads, which resembled palm trees, were also highly successful. They ranged across all major continents, possibly contributing to the diets of the plant-eating dinosaurs. The ginkgo, of which the maidenhair tree in eastern China is the only living relative, might have been the oldest genus of seed plants. Also dominating the landscape were conifers up to 5 feet across and 100 feet tall. Their petrified trunks are especially plentiful at Yellowstone National Park (Fig. 163).

About 110 million years ago, vegetation in the early Cretaceous underwent a radical change with the introduction of the angiosperms, flowering plants that evolved alongside pollinating insects. The plants offered pollinators, such as honeybees and birds, brightly colored and scented flowers, and sweet nectar. The unwary intruder was dusted with pollen, which it transported to the next flower it visited for pollination. Many angiosperms also depended on animals to spread their seeds, which were encased in tasty fruit that passed through the digestive tract and dropped some distance away.

The sudden appearance of angiosperms and their eventual domination over other plant life has remained a mystery. They might have originally exploited the weedy rift valleys that formed when Pangaea broke apart. The earliest angiosperms living in the lush rift valleys appear to have been large plants, growing as tall as magnolia trees. However, fossils discovered in Australia suggest that the first angiosperms there and perhaps elsewhere were small, herblike plants. Within a few million years after their introduction, the efficient flowering plants crowded out the once abundant ferns and gymnosperms. Angiosperms possessed water-conducting cells called vessel

Figure 163 *The three most prominent petrified tree stumps on North Scarp of Specimen Ridge in Yellowstone National Park, Wyoming.*

(Courtesy National Park Service)

elements that enabled the advanced plants to cope with drought conditions. Before such vessels appeared, plants were restricted to moist areas such as the wet undergrowth of rain forests.

The angiosperms were distributed worldwide by the end of the Cretaceous. Today, they include about a quarter-million species of trees, shrubs, grasses, and herbs. All major groups of modern plants (Fig. 164) were represented by the early Tertiary. The angiosperms dominated the plant world, and all modern families had evolved by about 25 million years ago. Grasses were the most significant angiosperms, providing food for hoofed mammals called ungulates. Their grazing habits evolved in response to the widespread availability of grasslands, which sparked the evolution of large herbivorous mammals and ferocious carnivores to prey on them.

Near the end of the Cretaceous, forests extended into the polar regions far beyond the present tree line. The most remarkable example is a well-preserved fossil forest on Alexander Island, Antarctica. To survive the harsh conditions, trees had to develop a means of protection against the cold since plants are more sensitive to the lack of heat than the absence of sunlight. They probably adapted mechanisms for intercepting the maximum amount of sunlight during a period when global temperatures were considerably warmer than today.

Figure 164 *A fossil leaf from the Raton Formation near Trinidad, Colorado.*

(Photo by W. T. Lee, courtesy USGS)

The cone-bearing plants prominent during the entire Mesozoic occupied only a secondary role during the Cenozoic. Tropical vegetation that was widespread during the Mesozoic withdrew to narrow regions around the equator in response to a colder, drier climate, resulting from a general uplift of the continents and the draining of the interior seas. Forests of giant hardwood trees that grew as far north as Montana are now occupied by scraggly conifers, a further indication of a cooler climate.

The rise of the angiosperms might have even contributed to the death of the dinosaurs and certain marine species at the end of the Cretaceous. By absorbing large quantities of carbon dioxide from the atmosphere, angiosperms caused a drop in global temperatures. Forests of broad-leaf trees and shrubs that were a favorite food of the dinosaurs apparently disappeared just prior to the ending of the Cretaceous. This therefore also contributed to the dinosaurs' downfall.

THE LARAMIDE OROGENY

Beginning about 80 million years ago, a large part of western North America uplifted. The entire Rocky Mountain Region (Fig. 165) from northern Mexico into Canada rose nearly a mile above sea level. This mountain building episode called the Laramide orogeny resulted from the subduction of oceanic crust beneath the West Coast of North America, causing an increase in crustal buoyancy. The Canadian Rockies consist of slices of sedimentary rock that were successively detached from the underlying basement rock and thrust eastward on top of each other. A region between the Sierra Nevada and the southern Rockies took a spurt of uplift during the past 20 million years, raising the area more than 3,000 feet.

Since the late Cambrian, the future Rocky Mountain region was near sea level. Farther west within about 400 miles of the coast, a mountain belt comparable to the present Andes formed above a subduction zone during the 80 million years prior to the Laramide. It was apparently responsible for the Cretaceous Sevier orogeny that created the Overthrust Belt in Utah and Nevada.

Figure 165 *The Rocky Mountains, looking southwest over Loveland Pass, north of South Park, Colorado.*

(Photo by T. S. Lovering, courtesy USGS)

A region from eastern Utah to the Texas panhandle that deformed during the late Paleozoic Ancestral Rockies orogeny was completely eroded by the time of the Laramide. The Rocky Mountain foreland region subsided as much as 2 miles between 85 million and 65 million years ago and then rose well above sea level, acquiring its present elevation around 30 million years ago.

To the west of the Rockies, numerous parallel faults sliced through the Basin and Range Province between the Sierra Nevada of California and the Wasatch Mountains of Utah. This resulted in a series of 20 north–south trending fault block mountain ranges. The Basin and Range covers southern Oregon, Nevada, western Utah, southeastern California, and southern Arizona and New Mexico. The crust bounded by faults is literally broken into hundreds of steeply tilted blocks and raised nearly a mile above the basin, forming nearly parallel mountain ranges up to 50 miles long.

Death Valley (Fig. 166), at 280 feet below sea level, is the lowest place on the North American continent. The region was originally elevated several thousand feet higher during the Cretaceous. The area collapsed when the continental crust thinned from extensive block faulting, with one block of crust lying below another. The Great Basin area is a remnant of a broad belt of mountains and high plateaus that subsequently collapsed after the crust was pulled apart following the Laramide.

The rising Wasatch Range of north-central Utah (Fig. 167) is an excellent example of a north-trending series of faults, one below the other. The fault blocks extend for 80 miles, with a probable net slip along the west side of 18,000 feet. The Tetons of western Wyoming were upfaulted along the eastern flank and downfaulted to the west. The rest of the Rocky Mountains

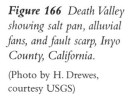

Figure 166 *Death Valley showing salt pan, alluvial fans, and fault scarp, Inyo County, California.*

(Photo by H. Drewes, courtesy USGS)

Figure 167 *The Wasatch Range of north-central Utah.*

(Photo by R. R. Woolley, courtesy USGS)

evolved by a process of upthrusting similar to the plate collision and subduction that raised the Andes of Central and South America. The Andes continue to rise due to an increase in crustal buoyancy caused by the subduction of the Nazca plate beneath the South American plate.

CRETACEOUS WARMING

During the Cretaceous, plants and animals were especially prolific and ranged practically from pole to pole. The deep ocean waters, which are now near freezing, were about 15 degrees Celsius during the Cretaceous. The average global surface temperature was 10 to 15 degrees warmer than at present. Conditions were also much warmer in the polar regions, with a temperature difference between the poles and the equator of only 20 degrees, or about half that of today.

The drifting of continents into warmer equatorial waters might have accounted for much of the mild climate during the Cretaceous. By the time of the initial breakup of the continents about 180 million years ago, the climate began to warm dramatically. The continents were flatter, the mountains were lower, and the sea levels were higher. Although the geography during this time was important, it did not account for all of the warming.

The movement of the continents was more rapid than today, with perhaps the most vigorous plate tectonics the world has ever known. About

120 million years ago, an extraordinary burst of submarine volcanism struck the Pacific Basin. It released vast amounts of gas-laden lava onto the ocean floor. The volcanic spasm is evidenced by a collection of massive undersea lava plateaus that formed almost simultaneously, the largest of which, the Ontong Java, is about two-thirds the size of Australia. It contains at least 9 million cubic miles of basalt, enough to bury the entire United States under 3 miles of lava.

The surge of volcanism increased the production of oceanic crust as much as 50 percent. This rise in volcanic activity provided perhaps the greatest contribution to the warming of Earth by producing 4 to 8 times the present amount of atmospheric carbon dioxide. As a result, worldwide temperatures averaged 7.5 to 12.5 degrees higher than today.

For the next 40 million years, Earth's geomagnetic field, which normally reverses polarity quite often on a geologic time scale (Table 10), stabilized. It assumed a constant orientation due to several mantle plumes that produced tremendous basaltic eruptions. This greater volcanic activity increased the carbon dioxide content of the atmosphere, producing the warmest global climate in 500 million years. Carbon dioxide also provided an abundant source of carbon for green vegetation and contributed to its prodigious growth, supplying a substantial diet for herbivorous dinosaurs.

Polar forests extended into latitudes 85 degrees north and south of the equator, as indicated by fossilized remains of an ancient forest that thrived on the now frozen continent of Antarctica. Evidence of a warm climate that supported lush vegetation is provided by coal seams running through the Transantarctic

TABLE 10 COMPARISON OF MAGNETIC REVERSALS WITH OTHER PHENOMENA (IN MILLION YEARS AGO)

Magnetic Reversal	Unusual Cold	Meteorite Activity	Sea Level Drops	Mass Extinctions
0.7	0.7	0.7		
1.9	1.9	1.9		
2.0	2.0			
10				11
40			37–20	37
70			70–60	65
130			132–125	137
160			165–140	173

Mountains, which are among the most extensive coal beds in the world. Alligators and crocodiles lived in the high northern latitudes as far north as Labrador, whereas today they are confined to warm, tropical areas. The duck-billed hadrasaurs also lived in the Arctic and Antarctic regions.

The positions of the continents might have contributed to the warming of the climate during most of the Mesozoic. Continents bunched together near the equator during the Cretaceous allowed warm ocean currents to carry heat poleward. High-latitude oceans are less reflective than land. So they absorbed more heat, further moderating the climate.

THE INLAND SEAS

In the late Cretaceous and early Tertiary, land areas were inundated by the ocean, which flooded continental margins and formed great inland seas. Some of these split continents in two. Seas divided North America in the Rocky Mountain and high plains regions. South America was cut in half in the region that later became the Amazon basin. Eurasia was split by the joining of the Tethys Sea and the newly formed Arctic Ocean.

The oceans of the Cretaceous were also interconnected in the equatorial regions by the Tethys and Central American Seaways (Fig. 168). They provided a unique circumglobal oceanic current system that made the climate equable. Mountains were lower and sea levels higher. The total land surface declined to perhaps half its present size. The Appalachians, which were an imposing mountain range at the beginning of the Triassic, were eroded down to stumps in the Cretaceous. Erosion toppled the once towering mountain ranges of Eurasia as well.

Great deposits of limestone and chalk were laid down in Europe and Asia, which is how the Cretaceous received its name. Seas invaded Asia, Africa, Australia, South America, and the interior of North America. About 80 million years ago, the Western Interior Cretaceous Seaway (Fig. 169) was a shallow body of water that divided the North American continent into the western highlands and the eastern uplands. The western highlands comprised the newly forming Rocky Mountains and isolated volcanoes. The eastern uplands consisted of the Appalachian Mountains.

Eastward of the rising Rocky Mountains was a broad coastal plain composed of thick layers of sediments eroded from the mountainous regions and extending to the western shore of the interior seaway. These sediment layers were later lithified and upraised. Today they are exposed as impressive cliffs in the western United States (Fig. 170). Along the coast and extending some distance inland were extensive wetlands, where dense vegetation grew in the subtropical climate. Inhabiting these areas were fish, amphibians, aquatic turtles,

Figure 168 *Distribution of continents around the Tethys Sea during the late Cretaceous.*

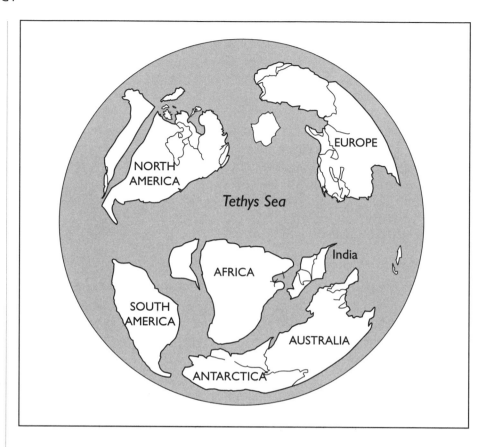

crocodiles, and primitive mammals. The dinosaurs included herbivorous hadrosaurs and triceratops along with the carnosaurs that preyed on them.

Toward the end of the Cretaceous, North America and Europe were no longer in contact, except for a land bridge that spanned Greenland to the north. The strait between Alaska and Asia narrowed, creating the practically landlocked Arctic Ocean. The South Atlantic continued to widen, with South America and Africa separated by more than 1,500 miles of ocean. Africa moved northward and began to close the Tethys Sea, leaving behind Antarctica, which was still joined to Australia.

As Antarctica and Australia continued to move eastward, a rift developed that eventually separated them. After rifting apart, Australia moved into the lower latitudes, while Antarctica drifted into the southern polar region and accumulated a massive ice sheet. Meanwhile, the northward-drifting subcontinent of India narrowed the gap between itself and southern Asia at a rate of about 2 inches per year. During its journey after the breakup with Gondwana, no known mammals appear to have existed in India until after its collision with Eurasia.

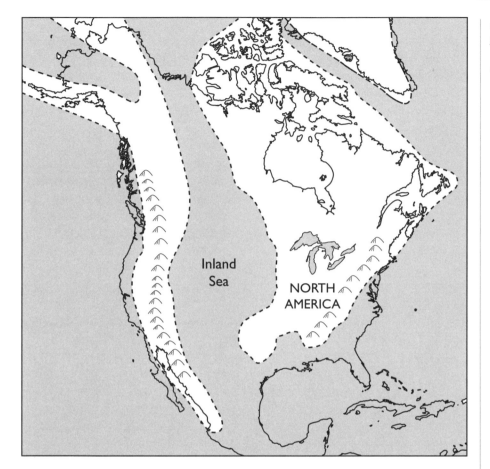

Inland
Sea

NORTH
AMERICA

Figure 170 *Cretaceous Mancos shale and Mesa Verde Formation, Mesa Verde National Park, Montezuma County, Colorado.*

(Photo by L. C. Huff, courtesy USGS)

229

Apparently during the middle Cretaceous, Australia, while still attached to Antarctica, wandered near the Antarctic Circle and acquired a thick mantle of ice. Large, out-of-place boulders strewn across the great central desert suggest that ice existed there even during the warm Cretaceous about 100 million years ago. At this time, Australia was still attached to Antarctica and straddled the Antarctic Circle, inside of which winters were sunless and cold. Maintaining wintertime temperatures above freezing in the interiors of large continents at high latitudes is difficult because they do not receive warmth from the ocean.

Like most continents of the Cretaceous, the interior of Australia contained a large inland sea. The continents were generally flatter, and sea levels were hundreds of feet higher than today. Sediments settling on the floor of the basin lithified into sandstone and shale. They were later exposed when the land uplifted and the sea departed at the end of the period. After Australia had drifted into the subtropics, the central portion of the continent became a large desert. Lying in the middle of the sedimentary deposits are curious-looking boulders of exotic rock called dropstones. They measure as much as 10 feet across and came from a great distance away. Rivers or mudslides could not have carried the boulders into the middle of the basin because such torrents would have disturbed the smooth sediments composed of fine-grained sandstone and shale.

The appearance of these strange boulders out in the middle of the desert suggests they rafted out to sea on slabs of drift ice. When the ice melted, the huge rocks simply dropped to the ocean floor, where their impacts disturbed the underlying sediment layers. The boulders apparently were not dropped by permanent glaciers but by seasonal ice packs that formed in winter. During the cold winters, portions of the interior coastline froze into pack ice. Rivers of broken ice then flowed into the inland sea, carrying with them embedded boulders dropped more than 60 miles from shore.

Evidence of ice rafting of boulders during the Cretaceous also exists in glacial soils in other areas of the world such as the Canadian Arctic and Siberia. This suggests that the high latitudes still had cold climates, in which ice formed easily even during one of the warmest periods in Earth history. Boulders were also found in sediments from other warm periods as well. The same ice-rafting process is occurring even today in the Hudson Bay.

When the Cretaceous ended, the seas regressed from the land due to lowering sea levels, and the climate grew colder. The last stage of the Cretaceous, called the Maestrichtian, was the coldest interval of the period. The decreasing global temperatures and increasing seasonal variation in the weather made the world stormier, with powerful gusty winds that wreaked havoc over Earth.

No clear evidence for significant glaciation during this time has been found. Yet most warmth-loving species, especially many of those living in the Tethys Sea, disappeared when the Cretaceous came to an end. The extinctions appear to have been gradual, occurring over a period of 1 to 2 million years.

Moreover, those species already in decline, including the dinosaurs and pterosaurs, might have been dealt a final death blow from above.

AN ASTEROID IMPACT

At the end of the Cretaceous, the dinosaurs along with 70 percent of all known species became extinct. Because the division between the Cretaceous and the Tertiary, called the K–T boundary, is not a sharp break but might represent up to 1 million years or more, this extinction was not necessarily sudden. It could have taken place over an extended period. Many dinosaurs along with other species were already in decline several million years before the end of the Cretaceous. Triceratops (Fig. 171), whose vast herds covered the entire globe and might have contributed to the decline of other dinosaur species, were among the last dinosaurs to die.

One theory attempts to explain the extinction of the dinosaurs and more than 70 percent of other species at the end of the Cretaceous. One or more large asteroids or comets struck Earth with an explosive force equivalent to 100 trillion tons of dynamite, which is equal to 1 million eruptions of Mount St. Helens. Sphereule layers up to 3 feet thick found in the Gulf of Mexico are related to the 65-million-year-old Chicxulub impact structure off the Yucatan Peninsula of Mexico. The spherules resemble the glassy chondrules (rounded granules) in carbonaceous chondrites, which are carbon-rich meteorites, and in lunar soils.

Figure 171 *Vast herds of triceratops roamed all parts of the world toward the end of the Cretaceous.*

231

The impact sent 500 billion tons of debris into the atmosphere. Glowing hot bits of crater debris flying past and back through the atmosphere set ablaze global-wide forest fires upon landing. The raging inferno burned perhaps one-quarter of all vegetation on the continents, turning a large part of Earth into a smoldering cinder. A heavy blanket of dust and soot encircled the entire globe and lingered for months, cooling the planet and halting photosynthesis. The planet was plunged into environmental calamity.

A catastrophe on this scale would have destroyed most terrestrial habitats and caused extinctions of tragic proportions. Generally, no animal heavier than 50 pounds survived the extinction. A large body size appears to have been a severe disadvantage among terrestrial animals. Species living in the tropics that relied on steady warmth and sunshine, such as the coral reef communities, were especially hard hit. For example, the rudists, which built reeflike structures, completely died out along with half of all bivalve genera.

A massive bombardment of meteorites might also have stripped away the upper atmospheric ozone layer, bathing Earth in the Sun's deadly ultraviolet rays. The increased radiation would have killed land plants and animals as well as primary producers in the surface waters of the ocean. The mammals, which were no larger than rodents, coexisted with the dinosaurs for more than 100 million years. However, because they were mostly nocturnal and remained in their underground borrows in the daylight hours, coming out only at night to feed, the mammals would have been spared from the onslaught of ultraviolet radiation during the daytime.

In the aftermath of the bombardment, Earth would have succumbed to a year of darkness under a thick brown smog of nitrogen oxide. Surface waters, poisoned by trace metals leached from the soil and rock, and global rains as corrosive as battery acid would have dealt a deadly blow to terrestrial lifeforms. Plants that survived as seeds and roots would have been relatively unscathed. The high acidity levels would have dissolved the calcium carbonate shells of marine organisms, while those with silica shells would have survived as they have done during other crises. Land animals living in burrows would have been well protected. Creatures living in lakes buffered against the acid would have survived the meteorite impact quite well.

The impact could also have caused widespread extinctions of microscopic marine plants called calcareous nannoplankton, which produce a sulfur compound that aids in cloud formation. With the death of these creatures, cloud cover would have decreased dramatically, triggering a global heat wave extreme enough to kill off the dinosaurs and most marine species. This contention is supported by the fossil record. It shows that ocean temperatures rose 5 to 10 degrees Celcius for tens of thousands of years beyond the end of the Cretaceous. During this time, over a period of almost half a million years, more than 90 percent of the calcareous nanno-

Figure 172 *The southwest slope of South Table Mountain, Golden, Colorado. The boundary between the Cretaceous and Tertiary periods lies 10 feet below where the man is standing.*

(Photo by R. W. Brown, courtesy USGS)

plankton disappeared along with most marine life in the upper portions of the ocean.

Sixty-five-million-year-old sediments found at the boundary between the Cretaceous and Tertiary periods throughout the world (Fig. 172) contain shocked quartz grains with distinctive lamellae, common soot from global forest fires, rare amino acids known to exist only on meteorites, the mineral stishovite—a dense form of silica found nowhere except at known impact sites, and unique concentrations of iridium—a rare isotope of platinum relatively abundant on meteorites and comets but practically nonexistent in Earth's crust.

The geologic record holds clues to other giant meteorite impacts with anomalous iridium concentrations that coincide with extinction episodes. However, they are not nearly as intense as the iridium concentrations in beds marking the end of the Cretaceous, which are as high as 1,000 times background levels. This suggests that the end–Cretaceous extinction might have been a unique event in the history of life on Earth.

After discussing life living in the warm Cretaceous, the next chapter will examine the evolution of the mammals and the geology of the Tertiary period.

13

TERTIARY MAMMALS
THE AGE OF ADVANCED SPECIES

This chapter examines the evolution of the mammals and the changing landscapes during the Tertiary period. The Cenozoic era, beginning 65 million years ago, is synonymous with the "age of mammals." Because of their great diversity, many more species of plants and animals are alive today than at any other time in geologic history. The appearance of the grasses early in the period spawned the evolution of ungulates or hoofed animals and voracious carnivores to prey on them. The prosimians (pre-apes) were also on the scene and gave rise to the anthropoids, the ancestors of apes and humans.

Extremes in climate and topography created a greater variety of living conditions than at any other equivalent span of geologic time. The rigorous environments presented many challenging opportunities for plants and animals. The extent to which they invaded diverse habitats was truly remarkable. This was a time of continuous change. All species had to adapt to a wide range of living conditions. The changing climate patterns resulted from the movement of continents toward their present positions and from the intense mountain building that raised most ranges of the world.

THE MAMMALIAN ERA

The first mammals were tiny, shrewlike creatures that appeared in the late Triassic about 220 million years ago, at roughly the same time as the dinosaurs. The two groups coexisted for about 150 million years thereafter. Mammals evolved from bulky, cold-blooded creatures that were themselves descendants of the reptiles. Mammals descended from the mammal-like reptiles, which were later driven into extinction by the dinosaurs about 160 million years ago. The mammals then began to branch into new forms, following the breakup of Pangaea. The multituberculates were probably the most interesting group of mammals that ever lived. They evolved about the same time as the dinosaurs and became extinct a little more than 30 million years ago, long after the dinosaurs disappeared.

One of the oldest mammalian skeletons ever uncovered is that of a 140-million-year-old symmetrodont. It appears to be a missing link between egg-laying and therian (live-birth) mammals. A bizarre, rat-sized animal dating back at least 120 million years walked on mammalian front legs and splayed reptilian hind legs. The animal was a close relative to the common ancestor of all mammals alive today. Around 100 million years ago, mammals began to branch into novel forms as new environments became available after the breakup of the supercontinent Pangaea.

The early mammals (Fig. 173) developed over a period of more than 100 million years into the first therian mammals, the ancestors of all living marsupials and placentals. During this time, mammals progressed and became more adaptive in a terrestrial environment. Mammalian teeth evolved from simple cones that were replaced repeatedly to more complex forms replaced only once at maturity. However, the mammalian jaw and other parts of the skull still shared many similarities with reptiles. One mysterious group known as the triconodonts, ranging from 150 to 80 million years ago, were primitive protomammals. They were possible ancestors of the monotremes, represented today by the Australian platypus and echidna, which lay eggs and walk with a reptilian sprawling gait.

The ancient mammals were forced into a nocturnal lifestyle, requiring the evolution of highly acute senses along with an enlarged brain to process the information. The archaic mammals developed a fourfold increase in relative brain size compared with the reptiles. Thereafter, they achieved no substantial increase in brain size for at least 100 million years, indicating an adaptation to a lengthy, stable ecological niche during the Mesozoic.

Following the extinction of the dinosaurs 65 million years ago, mammals became the recipients of daytime niches along with a preponderance of new sensory signals for the brain to organize as they competed with each other in a challenging environment. Another fourfold increase in relative brain size occurred about 50 million years ago in response to adaptive radiation of mam-

Figure 173 *Fossil mammal skeletons on display at the Museum of Geology, South Dakota School of Mines at Rapid City.*

mals into new environments. During this time, rodents, the largest group of mammals, appeared in the fossil record. During the rest of the Cenozoic, mammalian brains gradually grew larger in proportion to their bodies. Intelligent activity is generally the key to mammalian success, implying a certain degree of freedom of action. With their superior brains, mammals could compete successfully with much stronger animals.

When the dinosaurs left the stage at the end of the Cretaceous, the mammals were waiting in the wings, poised to conquer Earth. Because the dinosaurs represented the largest group of animals, their departure left the world wide open to invasion by the mammals. After the extinction of the dinosaurs, mammals began to radiate into dazzling arrays of new species. In less than 10 million years, all 18 modern orders of mammals were established. In addition, all orders of hoofed mammals emerged in full bloom. The small, nocturnal mammals evolved into larger animals, some of which were evolutionary dead ends. Of the 30 or so orders of mammals that existed during the early Cenozoic, only half had lived in the proceeding Cretaceous while almost two-thirds are still living today.

The evolution of the mammals following the dinosaur extinction, however, was not gradual. It progressed in fits and starts. The early Tertiary was characterized by an evolutionary lag, as though the world had not yet awakened from the great extinction. By the end of the Paleocene epoch, about 54 million years ago, when temperatures were on the rise, mammals began to diversify rapidly. About 37 million years ago, a sharp extinction event took many archaic mammal species, which were large, peculiar looking animals (Fig. 174). Afterward, most of the truly modern mammals began to evolve.

The extinction coincided with changes in the deep-ocean circulation and eliminated many species of marine life in the shallow seas that flooded the European continent. The separation of Greenland from Europe might have allowed frigid Arctic waters to drain into the North Atlantic, significantly lowering its temperature and causing most types of foraminifers (marine protozoans) to disappear. The climate grew much colder and the seas withdrew from the land (Fig. 175), as the ocean dropped 1,000 feet to perhaps its lowest level of the last several hundred million years.

Much of the drop in sea level resulted from the accumulation of massive ice sheets atop Antarctica, which had drifted over the South Pole. A large fall in sea level due to a major expansion of the Antarctic ice sheet led to another extinction about 11 million years ago. These cooling events removed the most vulnerable of species. So those living today are more robust, having withstood

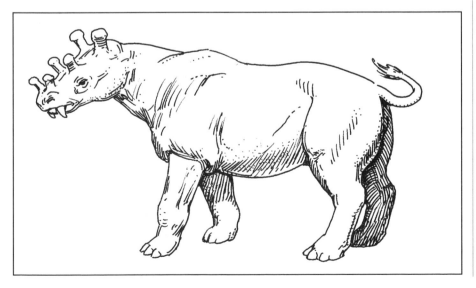

Figure 174 An extinct Eocene five-horned, saber-toothed, plant-eating mammal.

Figure 175
Paleogeography of the upper Tertiary in North America when inland seas withdrew from the continents.

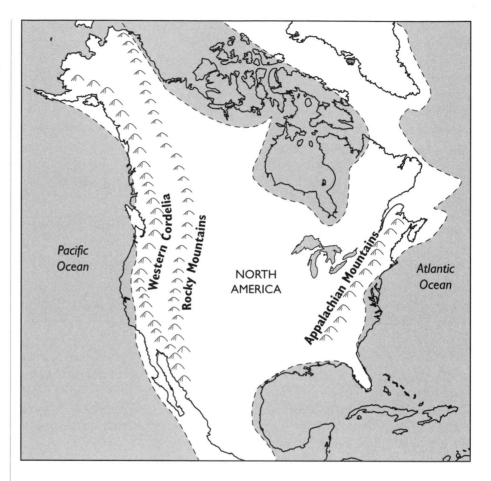

the extreme environmental swings over the last 3 million years, when glaciers spanned much of the Northern Hemisphere.

The number of mammalian genera rose to a high of 130 about 55 million years ago. Thereafter, the number of genera waxed and waned. It dropped to 60 and rose back up to 120 presumably in response to climate change and migration. These fluctuations lasted millions of years, yet diversity always converged on an equilibrium of about 90 genera. Greater speciation resulting in stiffer competition caused higher extinction rates, which maintained mammalian genera at a constant number.

Mammals are warm-blooded, which gives them a tremendous advantage. A stable body temperature finely tuned to operate within a narrow thermal range provides a high rate of metabolism independent of the outside temperature. Therefore, the work output of heart, lung, and leg muscles increases enormously, allowing mammals to out perform and out endure reptiles. Mammals also have a coat of insulation, comprising an outer layer of fat and fur to prevent the escape of body heat during cold weather.

Other distinguishable mammalian features include four-chambered hearts, a single bone in the lower jaw, highly differentiated teeth, and three small ear bones that migrated from the jawbone backward as the brain grew larger to improve hearing greatly. Mammals have live births. They possess mammary glands that provide a rich milk to suckle their young, which generally are born helpless. Mammals have the largest brains, capable of storing and retaining impressions. Therefore, they lived by their wits, which explains their great success. They conquered land, sea, and air and are established, if only seasonally, in all parts of the world.

The drifting of continents isolated many groups of mammals, and these evolved along independent lines. For example, Madagascar broke away from Africa about 125 million years ago. As such, it has none of the large mammals that live on the adjacent continent except the hippopotamus. The hippos mysteriously landed on the island after it had drifted some distance from the African mainland.

For the last 40 million years or so, Australia has been an island continent, without a land link to other landmasses. It is home to many strange egg-laying mammals called monotremes. These include the spiny anteater and platypus, which should rightfully be classified as surviving mammal-like reptiles. Marsupials are primitive mammals that suckle their tiny infants in belly pouches. They originated in North America around 100 million years ago, migrated to South America, crossed over to Antarctica when the two continents were still in contact, and landed in Australia before it broke away from Antarctica.

Today, 13 of the world's 16 marsupial families reside only in Australia. The Australian marsupials consist of kangaroos, wombats, and bandicoots, with opossums and related animals occupying other parts of the world. The largest marsupial fossil found is that of diprotodon (Fig. 176), which was about the size of a rhinoceros. Many large marsupials, including giant kangaroos, disappeared soon after early humans invaded the continent some 60,000 years ago.

Camels, which originated about 25 million years ago, migrated out of North America to other parts of the world by connecting land bridges. Horses originated in western North America during the Eocene when they were only about the size of small dogs. As they became progressively larger, their faces and teeth grew longer as the animals switched from browsing to grazing, and their toes fused into hoofs. The giraffes shifted from grazing on grass to browsing on shoots, twigs, and leaves. Their necks lengthened in response to reaching the tall branches. Many types of hoofed animals, called ungulates, evolved in response to increasing grasslands throughout the world.

All major groups of modern plants were represented in the early Tertiary (Fig. 177). The angiosperms dominated the plant world. All modern families

239

Figure 176 *Diprotodon was the world's largest marsupial.*

Figure 177 *Tertiary plants of the Chickaloon Formation, Cook Inlet region, Alaska.*

(Photo by J. A. Wolfe, courtesy USGS)

appear to have evolved by about 25 million years ago. Grasses were the most important angiosperms, providing food for ungulates throughout the Cenozoic. The grazing habits of many large mammals likely evolved in response to the widespread availability of grasslands.

The first primates lived some 60 million years ago and were about the size of a mouse. Then, the primate family tree split into two branches. Monkeys were on one limb. The great apes, including the hominoids—our humanlike ancestors, were on the other. Beginning about 37 million years ago, the New World monkeys unexplainably migrated from Africa to South America when those continents had already drifted far apart. About 30 million years ago, the precursors of apes lived in the dense tropical rain forests of Egypt, which is now mostly desert. These apelike ancestors migrated out of Africa and entered Europe and Asia between about 25 and 10 million years ago.

Between 12 and 9 million years ago, the forests of Europe were home to a tree-living, fruit-eating ape called *Dryopithecus,* which is thought to have evolved into *Ramapithecus,* an early Asian hominoid that had more advanced characteristics than earlier species. Between 9 and 4 million years ago, the fossil record jumps from the hominid-like but mainly ape form of *Ramapithecus* to the true hominids and the human line of ascension. During this time, much of Africa entered a period of cooler, drier climate, which caused the forests to retreat, offering many evolutionary challenges to our human ancestors.

MARINE MAMMALS

Some 70 species of marine mammals known as cetaceans were among the most adaptable animals and included dolphins, porpoises, and whales, which evolved during the middle Cenozoic. The dolphins had reached the level of intelligence comparable to living species by 20 million years ago probably due to the stability of their ocean environment. Sea otters, seals, walruses, and manatees are not fully adapted to a continuous life at sea and have retained many of their terrestrial characteristics. The manatees, which have inhabited Florida waters for 45 million years, are rapidly becoming an endangered species.

Pinnipeds, meaning "fin-footed," are a group of marine mammals with four flippers. The three surviving forms include seals, sea lions, and walruses. The true seals without ears are thought to have evolved from weasel-like or otterlike forms. Sea lions and walruses, however, are believed to have developed from bearlike forms. This type of dual development, called diphyletic evolution, makes originally dissimilar animals resemble each other simply because they have adapted to a life in the same environment, in this case water. The similarity in their flippers, however, suggests that all pinnipeds evolved from a single land-based mammal that entered the sea millions of years ago.

Whales presented a mystery as vast as the animals themselves. The ancestor of modern whales appears to have been a four-legged carnivorous mammal that walked on land and swam in rivers and lakes about 57 million years ago. The legless leviathans evolved from mammals known as ungulates, whose best-known characteristic is a set of hoofed feet. Whales adapted to swimming, diving, and feeding that matches or surpasses fish and sharks. They might have gone through a seal-like amphibious stage early in their evolution.

Today, their closest relatives are the artiodactyls, or hoofed mammals with an even number of toes, such as cows, pigs, deer, camels, giraffes, and hippos. However, exactly where whales fit in the tree of hoofed mammals remains controversial. Genetic evidence suggests that whales and hippopotamuses are closely related. Both groups share particular aquatic adaptations, such as the ability to nurse their young and communicate underwater. The ancestor of both whales and hippos might have ventured into the sea as early as 55 million years ago.

The earliest whales had limbs that would have been clumsy on land. However, their large feet and flexible spines allowed them to undulate their backs up and down to propel themselves through the water like modern whales. Today, whales have only vestiges of leg bones and lack ankle bones. The first whales probably lived in freshwater before entering the sea and did not stray far from the coastline because they needed to return to a river to drink. Ancestors of the giant blue whale (Fig. 178), the largest animal on Earth, even dwarfing the biggest dinosaurs that ever lived, evolved from ancient toothed whales about 40 million years ago.

TERTIARY VOLCANICS

Volcanic activity was extensive during the Tertiary. Massive flood basalts were caused by hot-spot volcanism, as plumes of magma rose to the surface from deep within the mantle. India's Deccan Traps (Fig. 179) were among the greatest outpourings of basalt on land during the last 250 million years. About 65 million years ago, a giant rift ran down the west side of India. Huge volumes of molten magma poured onto the surface. Some 100 separate flows spilled more than 350,000 cubic miles of lava onto much of west-central India, totaling up to 8,000 feet thick over a period of several million years. If spread evenly around the world, that vast amount of lava would envelop the entire Earth in a layer of volcanic rock some 10 feet thick.

During the eruptions, India was about 300 miles northeast of Madagascar, as it continued drifting toward southern Asia. The Seychelles Bank is a large oceanic volcanic plateau that became separated from the Indian subcontinent and is now exposed on the surface as several islands. The Ninety East

Figure 178 *Blue whales are the largest animals on Earth.*

Ridge is an undersea volcanic mountain range that runs 3,000 miles south from the Bay of Bengal, India. It formed when the Indian plate passed over a hot spot as it continued drifting toward Asia. The massive outpourings of carbon dioxide–laden lava might have created the extraordinary warm climate of the Paleocene that sparked the evolution of the mammals.

Continental rifting contemporary with the Deccan Traps eruptions began separating Greenland from Norway and North America. The rifting poured out great flood basalts across eastern Greenland, northwestern Britain, northern Ireland, and the Faeroe Islands between Britain and Iceland. The island of ice is itself an expression of the Mid-Atlantic Ridge, where massive floods of basalt formed a huge volcanic plateau that rose above sea level about 16 million years ago.

Evidence of substantial explosive volcanism lies in an extensive region from the South Atlantic to Antarctica. The Kerguelen Plateau located north of Antarctica is the world's largest submerged volcanic plateau. It originated from

Figure 179 *The Deccan Traps flood basalts in India.*

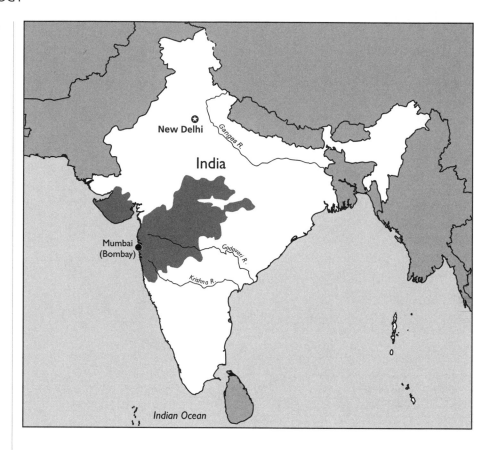

the ocean floor more than 90 million years ago, when a series of volcanic eruptions released immense quantities of basalt onto the Antarctic plate. The timing also coincided with a mass extinction of species.

Prior to the opening of the Red Sea and Gulf of Aden, massive floods of basalt covered some 300,000 square miles of Ethiopia, beginning about 35 million years ago. The East African Rift Valley extends from the shores of Mozambique to the Red Sea, where it splits to form the Afar Triangle in Ethiopia. For the past 25 to 30 million years, the Afar Triangle has been stewing with volcanism. An expanding mass of molten magma lying just beneath the crust uplifted much of the area thousands of feet.

In North America, major episodes of basalt volcanism occurred in the Columbia River Plateau, the Colorado Plateau, and the Sierra Madre region. A band of volcanoes stretching from Colorado to Nevada produced a series of very violent eruptions between 30 million and 26 million years ago. Beginning about 17 million years ago and extending over a period of 2 million years, great outpourings of basalt covered Washington, Oregon, and Idaho, creating the Columbia River Plateau (Fig. 180). Massive floods of lava enveloped an area of about 200,000 square miles, in places reaching 10,000 feet thick. Periodically, volcanic eruptions spewed out batches of basalt as large as 1,200 cubic miles, forming lava lakes up to 450 miles wide in a matter of days.

The volcanic episodes might be related to the Yellowstone hot spot, which was then positioned beneath the Columbia River Plateau region. The hot spot moved eastward relative to the North American plate. It can be traced

Figure 180 *Palouse Falls in Columbia River basalt, Franklin-Whitman Counties, Washington.*

(Photo by F. O. Jones, courtesy USGS)

by following volcanic rocks for 400 miles across Idaho's Snake River Plain. During the last 2 million years, it was responsible for three major episodes of volcanic activity in the vicinity of Yellowstone National Park in Wyoming, which are counted among the greatest catastrophes of nature.

CENOZOIC MOUNTAIN BUILDING

The Cenozoic is known for its intense mountain building. The spurt in mountain growth over the past 5 million years might have triggered the Pleistocene ice ages. The Rocky Mountains (Fig. 181), extending from Mexico to Canada, heaved upward during the Laramide orogeny from about 80 million to 40 million years ago. During the Miocene, beginning some 25 million years ago, a large portion of western North America uplifted. The entire Rocky Mountain region rose about a mile above sea level. Great blocks of granite soared high above the surrounding terrain. To the west in the Basin and Range Province, the crust stretched and thinned out, in some places dropping below sea level.

Arizona's Grand Canyon (Figs. 182 and 183) lies at the southwest end of the Colorado plateau, a generally mountain-free expanse that stretches from

Figure 181 *Granitic dikes cutting through horizontal sedimentary rocks, west Spanish Peak, Colorado.*

(Photo by G. W. Stose, courtesy USGS)

Figure 182 *A view of the Grand Canyon from Mojave Point.*

(Photo courtesy National Park Service)

Arizona north into Utah and east into Colorado and New Mexico. Initially, the area surrounding the canyon was practically flat. Over the last 2 billion years, heat and pressure buckled the land into mountains that were later flattened by erosion. Again, mountains formed and eroded, and the region flooded with shallow seas. The land was uplifted another time during the rising of the Rocky Mountains. Between 10 and 20 million years ago, the Colorado River began eroding layers of sediment, exposing the raw basement rock below. Its present course is less than 6 million years old. Substantial portions of the eastern Grand Canyon are geologically young, having been eroded within only the past million years.

About 30 million years ago, the North American continent approached the East Pacific Rise spreading center, the counterpart of the Mid-Atlantic Ridge. The first portion of the continent to override the axis of seafloor spreading was the coast of southern California and northwest Mexico. When the rift system and subduction zone converged, the intervening oceanic plate dove into a deep trench. The sediments in the trench were compressed and thrust upward to form California's Coast Ranges. A system of faults associated with the 650-mile-long San Andreas Fault (Fig. 184) crisscross the mountain belt. The Sierra Nevada to the east rose about 7,000 feet over the

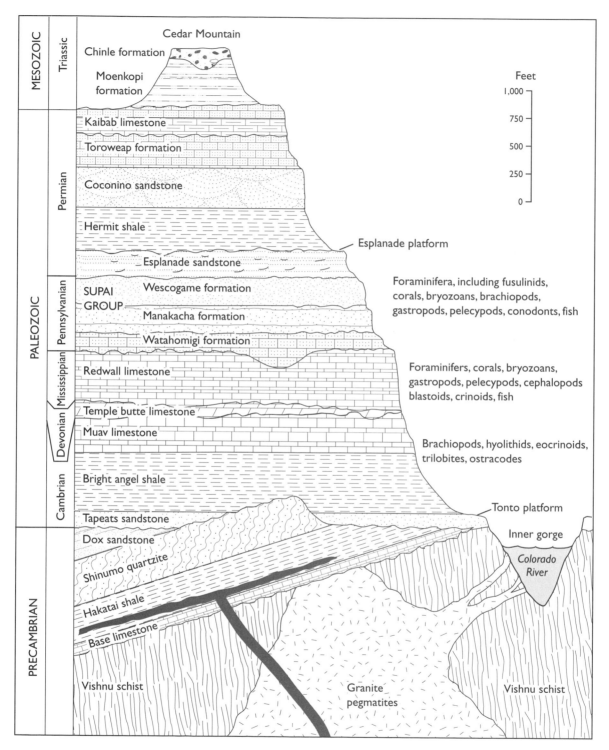

Figure 183 *A cross section of the Grand Canyon.*

last 10 million years and might be buoyed by a mass of hot rock in the upper mantle.

In the Pacific Northwest of the United States and British Columbia, the Juan de Fuca plate dove into the Cascadia subduction zone located beneath the continent. As the 50-mile-thick crustal plate subducted into the mantle, Earth's interior heat melted parts of the descending plate and the adjacent lithospheric plate, forming pockets of magma. The magma rose toward the surface, forming the volcanoes of the Cascade Range (Fig. 185), which erupted in one great profusion after another.

India and the rocks that comprise the Himalayas broke away from Gondwana early in the Cretaceous, sped across the ancestral Indian Ocean, and slammed into southern Asia about 45 million years ago. As the Indian and Asian plates collided, the oceanic lithosphere between them thrust under Tibet, destroying 6,000 miles of subducting plate. The increased buoy-

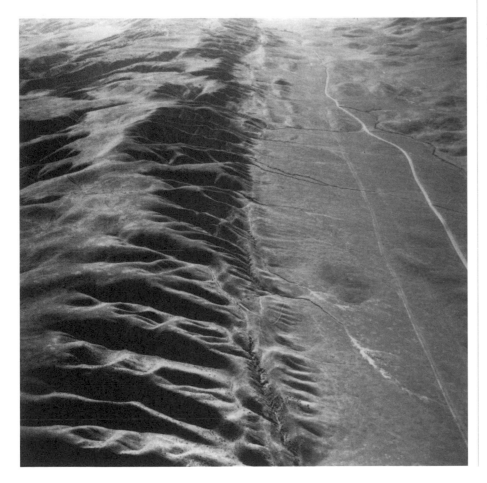

Figure 184 *The San Andreas Fault in Choice Valley, San Luis Obispo County, California.*

(Photo by R. E. Wallace, courtesy USGS)

Figure 185 *The south side of Mount Adams, with Mount Rainer in the background, Cascade Range, Yakima, County Washington.*

(Photo by A. Post, courtesy USGS)

ancy uplifted the Himalaya Mountains and the broad Tibetan Plateau (Fig. 186), the size of which has not been equaled on this planet for more than 1 billion years.

During the past 5 to 10 million years, the entire region rose over a mile in elevation. The continental collision heated vast amounts of carbonate rock, spewing several hundred trillion tons of carbon dioxide into the atmosphere. This might explain why Earth grew so warm during the Eocene epoch from 54 million to 37 million years ago, when temperatures reached the highest of the past 65 million years. According to the fossil record, winters were warm enough for crocodiles to roam as far north as Wyoming, and forests of palms, cycads, and ferns covered Montana.

About 50 million years ago, the Tethys Sea separating Eurasia from Africa narrowed as the two continents approached each other, then began to close off entirely some 20 million years ago. Thick sediments that had been accumulating for tens of millions of years buckled into long belts of mountain ranges on the northern and southern flanks (Fig. 187). The contact between the continents initiated a major mountain building episode that raised the Alps and other ranges in Europe and squeezed out the Tethys Sea.

This episode of mountain building, called the Alpine orogeny, raised the Pyrenees on the border between Spain and France, the Atlas Mountains of Northwest Africa, and the Carpathians in East-Central Europe. The Alps of northern Italy formed in much the same manner as the Himalayas, when the Italian prong of the African plate thrust into the European plate.

In South America, the mountainous spine that comprises the Andes running along the western edge of the continent rose throughout much of the Cenozoic due to an increase in crustal buoyancy from the subduction of

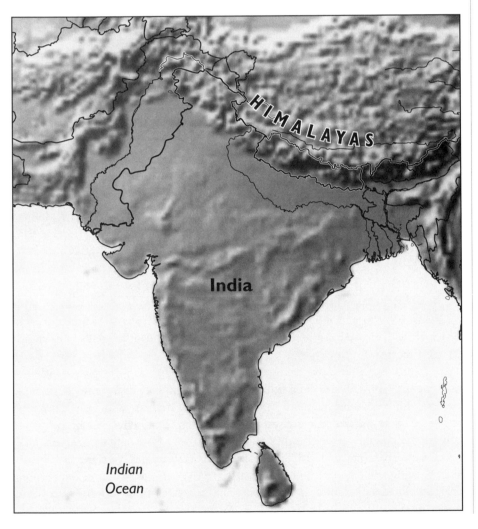

Figure 186 *The collision of the Indian subcontinent with Asia uplifted the Himalaya Mountains and the Tibetan Plateau.*

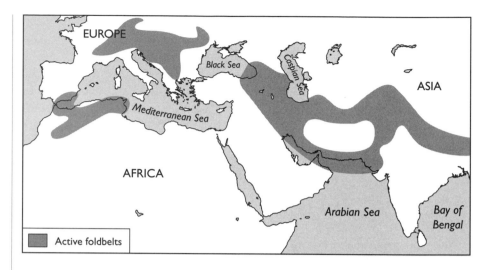

Figure 187 *Active fold belts in Eurasia, resulting from the collision of lithospheric plates.*

the Nazca plate beneath the South American plate. By the time all the continents had wandered to their present positions and all the mountain ranges had risen to their current heights, the world was ripe for the coming of the Ice Age.

TERTIARY TECTONICS

Changing climate patterns resulted from the movement of continents toward their present positions. Intense tectonic activity built landforms and raised most mountain ranges of the world. About 57 million years ago, Greenland began to separate from North America and Eurasia. Prior to about 8 million years ago, Greenland was largely ice free. However, today the world's largest island is buried under a sheet of ice up to 2 miles thick. Alaska connected with east Siberia and closed off the Arctic Basin from warm-water currents originating from the tropics, resulting in the formation of pack ice in the Arctic Ocean.

Except for a few land bridges exposed from time to time, plants and animals were prevented from migrating from one continent to another. A narrow, curved land bridge temporarily connected South America with Antarctica and assisted in the migration of marsupials to Australia. Sediments containing fossils of a large crocodile, a 6-foot flightless bird, and a 30-foot whale suggest that land bridges existed as late as 40 million years ago. Antarctica and Australia then broke away from South America and moved eastward. When the two continents rifted apart in the Eocene about 40 million years ago, Antarctica moved southward, while Australia continued moving in a northeastward direction. When Antarctica drifted over the South Pole, it acquired a permanent ice sheet that buried most of its terrain features (Fig. 188).

The Mid-Atlantic Ridge system, which generates new ocean crust in the Atlantic Basin, began to occupy its present location midway between the Americas and Eurasia/Africa about 16 million years ago. Iceland is a broad volcanic plateau of the Mid-Atlantic Ridge that rose above sea level. About 3 million years ago, the Panama Isthmus separating North and South America uplifted as oceanic plates collided. Prior to the continental collision, South America had been an island continent for the past 80 million years, during which time its mammals evolved undisturbed by outside competitors.

The barrier created by the land bridge isolated Atlantic and Pacific species. Extinctions impoverished the once rich fauna of the western Atlantic. The new landform halted the flow of cold water currents from the Atlantic into the Pacific. Along with the closing of the Arctic Ocean from

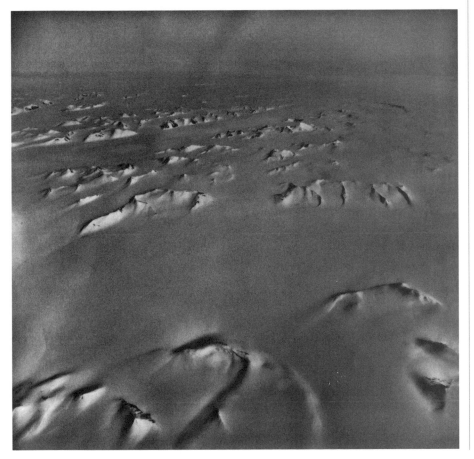

Figure 188 *The Antarctic Peninsula ice plateau, showing mountains literally buried in ice.*

(Photo by P. D. Rowley, courtesy USGS)

warm Pacific currents, this might have initiated the Pleistocene glacial epoch. Never before have permanent ice caps existed at both poles, suggesting that the planet has been steadily cooling since the Cretaceous.

CLOSING OF THE TETHYS

The Tethys Sea was a large, shallow equatorial body of water that linked the Indian and Atlantic Oceans. It separated the southern and northern continents during the Mesozoic and early Cenozoic. About 17 million years ago, the Tethys began to close off as Africa rammed into Eurasia, creating the Mediterranean, Black, Caspian, and Aral Seas (Fig. 189). The collision also initiated a major mountain building episode that raised the Alps and other ranges. The climates of Europe and Asia were warmer. Forests were more widespread and lusher than today.

The Mediterranean Basin was apparently cut off from the Atlantic Ocean 6 million years ago. An isthmus, created at Gibraltar by the northward movement of the African plate, formed a dam across the strait. Nearly 1 million cubic miles of seawater evaporated, almost completely emptying the basin over a period of about 1,000 years. The adjacent Black Sea, at 750 miles long and 7,000 feet deep, might have had a similar fate. Like the Mediterranean, it is a remnant of an ancient equatorial sea that separated Africa from Europe.

The collision of the African plate with the Eurasian plate squeezed out the Tethys. This resulted in a long chain of mountains and two major inland seas, the ancestral Mediterranean and a composite of the Black, Caspian, and

Figure 189 *The closing of the Tethys Sea by the collision of Africa with Europe and Asia about 20 million years ago, creating the Mediterranean, Black, Caspian, and Aral Seas.*

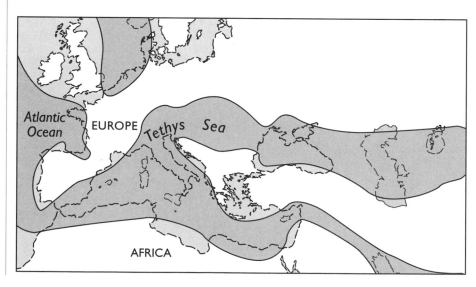

Aral Seas, called the Paratethys. It covered much of eastern Europe. About 15 million years ago, the Mediterranean separated from the Paratethys, which became a brackish (slightly salty) sea, much like the Black Sea of today.

The disintegration of the great inland waterway was closely associated with the sudden drying of the Mediterranean. The waters of the Black Sea drained into the desiccated basin of the Mediterranean. In a brief moment in geologic time, the Black Sea became practically a dry basin. Then during the last ice age, it refilled again and became a freshwater lake. The brackish, largely stagnant sea that occupies the basin today has evolved since the end of the last ice age.

After covering the evolution of the mammals during the Tertiary, the next chapter will examine the ice ages of the Quaternary period.

14

QUATERNARY
GLACIATION
THE AGE OF MODERN LIFE

This chapter examines the Ice Ages and the present interglacial during the Quaternary period. The Quaternary, from 3 million years ago to the present, witnessed a progression of ice ages that came and went almost like clockwork. The period is divided into the Pleistocene glacial epoch and the Holocene interglacial, which is also commensurate with the rise of civilization. Both the Tertiary and Quaternary are terms carried over from the old geologic time scale in which the Primary and Secondary periods represented ancient Earth history. The pronounced unequal lengths of the two periods is in recognition of the unique sequence of ice ages. Many geologists prefer to divide the Cenozoic into two nearly equal time intervals: the Paleogene from 65 to 26 million years ago and the Neogene from 26 million years ago to the present.

During the Pleistocene, the movement of continents to their present locations and the raising of land to higher elevations made geographic conditions ripe for a colder climate. Variations in the Earth's orbital motions might have provided the initial kick to trigger the growth of continental glaciers, which partly explains the recurrence of the ice age cycles about every

100,000 years. Once in place, the glaciers became self-sustaining by controlling the climate. Then, mysteriously in only a few thousand years, the great ice sheets collapsed and rapidly retreated to the poles. As the result of the Ice Ages, many northern lands owe their unusual topographies to massive ice sheets that swept down from the polar regions.

The Pleistocene, originally defined as a period of recent life based on the fossil record of modern organisms, has become synonymous with a period of glaciation that began some 3 million years ago. The Ice Age was so pervasive that ice sheets 2 miles or more thick enveloped upper North America and Eurasia, Antarctica, and parts of the Southern Hemisphere. In many areas, the glaciers stripped off entire layers of sediment down to the bare bedrock, erasing the entire geologic history of the region.

THE HUMAN ERA

About 3 million years ago, huge volcanic eruptions in the northern Pacific darkened the skies and global temperatures plummeted, culminating in a series of glacial episodes. The climate change prompted a shift from forested environments to open savanna habitats in Africa. These changing conditions produced many new animal species and spurred the evolution of early humans, who had to adapt to a new environment rapidly. Indeed, humans are products of the ice ages, which spanned the whole of human experience.

Humans direct ancestors—the hominids—evolved in Africa, probably from the same species that gave rise to the great apes, including the gorilla and chimpanzee. These species have 99 percent of the same genes as humans. Around 7 million years ago, much of Africa entered a period of cooler, drier climates when forests retreated and were replaced with grasslands. Life on the savanna, where humans' ancient ancestors roamed, was harsher and more challenging than life in the forests, where the apes lived. To survive under these difficult conditions, early humans rapidly evolved into intelligent, upright-walking species, whereas the apes are much the same today as they were millions of years ago.

An early hominid species called *Australopithecus* (Fig. 190) first appeared in Africa about 4 million years ago. It walked on two legs but retained many apelike features, such as long arms in relation to its legs and curved bones in its hands and feet. The species was quite muscular and considerably stronger than modern humans. Males stood a little less than 5 feet tall and weighed about 100 pounds. Females stood about 4 feet tall and weighed about 70 pounds. Two or more lines of australopithecine lived simultaneously in Africa and survived practically unchanged for more than 1 million years. After a lengthy period of apparent stability, all but one line became extinct, possibly due to a changing climate or habitat.

Figure 190 *The human ancestry probably began with* Australopithecus.

(Photo courtesy National Museums of Canada)

About 2.5 million years ago, a sudden change in climate spurred the evolution of the first humans. The drying of the African climate caused an increase in grasslands at the expense of forests, where human ancestors probably found food, shelter, and refuge from predators. With the stress of life on the open savannah, early humans might have developed larger brains, which gave rise to toolmaking skills and cooperative hunting techniques in order to survive.

The most successful of the early hominids was *Homo habilis,* which evolved a little more than 2 million years ago. It was a species in transition between primitive apelike hominids and early humans. Its brain was about half the size of that of modern humans. The limb bones were markedly different from earlier hominids and more closely resembled those of later humans. It was the first human species to make and use tools. It had well-developed speech centers, indicating a primitive language capability.

Homo habilis disappeared from Africa around 1.8 million years ago and was replaced by *Homo erectus,* the first human ancestor to have departed Africa. This species is widely accepted as human and appears to have evolved in Africa directly from *Homo habilis.* It could have also evolved independently in Asia and subsequently migrated to Africa. About 1 million years ago, this early human occupied southern and eastern Asia, where it lived until about 200,000 years ago. Its advanced features suggest a spurt of evolutionary development, as it shares many attributes of modern humans. Its brain was also larger, about two-thirds the size of a modern human's.

Many types of *Homo erectus* were scattered throughout the world. This suggests that anatomically modern humans evolved from this species in several places, possibly accounting for the differences in races among people today. Peking man, a variety of *Homo erectus* that lived in China about 400,000 years ago, dwelled in caves and was possibly the first to use fire. Another variety, called Java man, arrived in Java about 700,000 years ago. About 60,000 years ago, the descendants of Java man migrated to Australia and possibly other South Pacific islands.

The earliest *Homo sapiens* were called Cro-Magnon, named for the Cro-Magnon cave in France, where the first discoveries were made in 1868. They originated in Africa perhaps 200,000 years ago. Evidence also suggests they arose simultaneously in several parts of the world, as much as 1 million years ago, possibly evolving directly from *Homo erectus.* The Cro-Magnon shared most of the physical attributes of modern humans. The skull, whose brain case proportions were modern in structure, was short, high, and rounded. The lower jaw ended in a definite chin. The rest of the skeleton was slender and long limbed compared with earlier human species.

Sometime in the last ice age, Cro-Magnon appears to have advanced into Europe and Asia during a warm interlude when the climate was less severe. They probably lived much like present-day natives of the Arctic tundra,

Figure 191 *Arctic dwellers built houses of mammoth bones and tusks covered with hides.*

fishing the rivers and hunting reindeer and other animals. As with modern populations of Arctic peoples, human ancestors inhabiting higher latitudes were larger than those living near the equator. Due to the scarcity of wood in the cold tundra, ice age hunters on the central Russian plain built houses of mammoth bones and tusks covered with animal hides (Fig. 191). They burned bones and animal fat for heat and light.

The Neandertals were primitive *Homo sapiens,* named for the Neander Valley near Dusseldorf, Germany, where the first fossils were recognized in 1856. They are generally thought to have inhabited caves. However, they also occupied open-air sites, as evidenced by hearths and rings made of mammoth bones and stone tools normally associated with these people. During the last interglacial period, called the Eemian, from about 135,000 to 115,000 years ago, the Neandertals ranged over most of western Europe and central Asia, extending as far north as the Arctic Ocean.

Modern humans and Neandertals apparently coexisted in Eurasia for at least 60,000 years and shared many cultural advancements. Although European Neandertals living from 75,000 to 35,000 years ago were generally one-third larger than modern humans, their relative brain size was nearly equal. The Neandertals thrived in these regions until about 35,000 years ago, declining over a period of perhaps 5,000 years. The disappearance of the Neandertals might have resulted from the replacement or assimilation by modern humans.

THE PLEISTOCENE ICE AGES

For more than 200 million years from the end of the Permian to about 40 million years ago, no major ice caps covered the world. This suggests that the existence of ice sheets at both poles during the Pleistocene was a unique event in Earth history. Analysis of deep-sea sediments and glacial ice cores provides a historic record of the ice ages. They began about 3 million years ago, when a progression of glaciers sprawled across the northern continents.

During this time, the surface waters of the ocean cooled dramatically. Diatoms (Fig. 192), a species of algae with shells made of silica, sharply decline in the Antarctic. This presumably occurred when sea ice reached its maximum northern extent, thus shading the algae below. In the absence of sunlight for photosynthesis, the diatoms vanished. Their disappearance marks the initiation of the Pleistocene glacial epoch in the Northern Hemisphere.

The latest period of glaciation began about 115,000 years ago. It intensified about 75,000 years ago, possibly due to the massive Mount Toba eruption in Indonesia. It peaked about 18,000 years ago. During the height of the Ice Age, glaciers up to 2 or more miles thick enveloped Canada, Greenland, and Northern Europe (Fig. 193).

In North America, the largest ice sheet, called the Laurentide, blanketed an area of 5 million square miles. It extended from Hudson Bay, reaching northward into the Arctic Ocean and southward into eastern Canada, New England, and the upper midwestern United States. A smaller ice sheet, called the Cordilleran, originated in the Canadian Rockies. It engulfed western Canada and the northern and southern parts of Alaska. It left an ice-free corridor down the middle, used by people migrating from Asia into North America. Glaciers also invaded small portions of northwestern United States.

The largest ice sheet in Europe, called the Fennoscandian, fanned out from northern Scandinavia. It covered most of Great Britain as far south as London and large parts of northern Germany, Poland, and European Russia. A smaller ice sheet, called the Alpine, centered in the Swiss Alps, covered parts of Austria, Italy, France, and southern Germany. In Asia, ice sheets draped over the Himalayas and blanketed parts of Siberia.

In the Southern Hemisphere, only Antarctica had a major ice sheet. It expanded about 10 percent larger than its present size and extended as far as the tip of South America. Small ice sheets capped the mountains of Australia, New Zealand, and the Andes of South America. Elsewhere, alpine glaciers topped mountains that are presently ice free. Excess ice with nowhere to go except into the sea calved off to form icebergs. During the peak of the last ice age, icebergs covered half the area of the ocean. The ice floating in the sea reflected sunlight back into space, thereby maintaining a cool climate with average global temperatures about 5 degrees Celsius colder than today.

Some 10 million cubic miles of water were tied up in the continental ice sheets. They covered about one-third of the land surface, with an ice volume three times greater than its present size. The accumulated ice dropped the level of the ocean about 400 feet, advancing the shoreline tens of miles seaward. The drop in sea level exposed land bridges and linked continents, spurring the

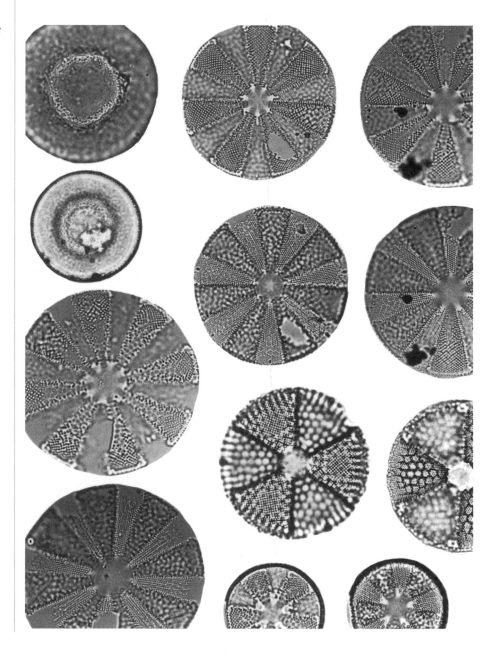

Figure 192 *Miocene-age diatoms from the Choptank Formation, Calvert County, Maryland.*

(Photo by G. W. Andrews, courtesy USGS)

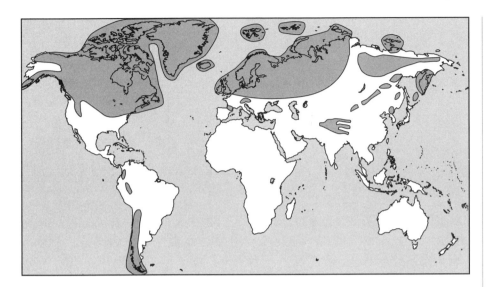

Figure 193 *The extent of the glaciers during the last ice age.*

migration of species, including humans, to various parts of the world. The great weight of the ice sheets caused the continental crust to sink deeper into the upper mantle. Even today, the northern lands are rebounding as much as half an inch per year long after the weight of the glaciers has been lifted.

The lower temperatures reduced the evaporation rate of seawater and lowered the average amount of precipitation, which caused the expansion of deserts in many parts of the world (Table 11). The fierce desert winds produced tremendous dust storms. The dense dust suspended in the atmosphere blocked sunlight, dropping temperatures even further. Most of the windblown sand deposits, called loess (Fig. 194), in the central United States were laid down during the Pleistocene Ice Ages.

The cold weather and approaching ice forced species to migrate to warmer latitudes. Ahead of the ice sheets, which advanced perhaps a few hundred feet per year, lush deciduous woodlands gave way to evergreen forests that yielded to grasslands. These grasslands became barren tundras and rugged periglacial regions on the margins of the ice sheets.

THE HOLOCENE INTERGLACIAL

Perhaps one of the most dramatic climate changes in geologic history took place during the present interglacial period known as the Holocene epoch. After some 100,000 years of gradual accumulation of snow and ice up to 2 miles and more thick, the glaciers melted away in only a few thousand years, retreating at a rate of several hundred feet annually. About one-third of the ice melted

TABLE 11   THE MAJOR DESERTS

Desert	Location	Type	Area (square miles x 1,000)
Sahara	North Africa	Tropical	3,500
Australian	Western/Interior	Tropical	1,300
Arabian	Arabian Peninsula	Tropical	1,000
Turkestan	Central Asia	Continental	750
North American	Southwest U.S./North Mexico	Continental	500
Patagonian	Argentina	Continental	260
Thar	India/Pakistan	Tropical	230
Kalahari	Southwest Africa	Littoral	220
Gobi	Mongolia/China	Continental	200
Takla Makan	Sinkiang, China	Continental	200
Iranian	Iran/Afganistan	Tropical	150
Atacama	Peru/Chile	Littoral	140

Figure 194 An exposure of loess standing in vertical cliffs, Warren County, Mississippi.

(Photo by E. W. Shaw, courtesy USGS)

Figure 195 *The Grand Coulee Dam on the Columbia River in northwest Washington, showing tortured scablands terrain.*

(Courtesy USGS)

between 16,000 and 12,000 years ago, when average global temperatures increased about 5 degrees Celsius to nearly present-day levels. A renewal of the deep-ocean circulation system, which was shut off or severely weakened during the Ice Age, might have thawed out the planet from the deep freeze.

A gigantic ice dam on the border between Idaho and Montana held back a huge lake hundreds of miles wide and up to 2,000 feet deep. Around 13,000 years ago, the sudden bursting of the dam sent waters gushing toward the Pacific Ocean. Along the way, the floodwaters carved out one of the strangest landscapes known, called the Channel Scablands (Fig. 195). Lake Agassiz, a vast reservoir of meltwater formed in a bedrock depression carved out by glaciers, sat at the edge of the retreating ice sheet in southern Manitoba, Canada.

When the North American ice sheet retreated, its meltwaters flowed down the Mississippi River into the Gulf of Mexico. After the ice sheet subsided beyond the Great Lakes, the meltwaters took an alternate route down the St. Lawrence River. The cold waters entered the North Atlantic Ocean. Simultaneously, the Niagara River Falls began cutting its gorge and has traversed more than 5 miles northward since the ice sheet melted.

The rapid melting of the glaciers culminated in the extinction of microscopic organisms called foraminifera (Fig. 196). They met their demise when a torrent of meltwater and icebergs spilled into the North Atlantic. The massive floods formed a cold freshwater lid on the ocean that significantly changed the salinity of the seawater. The cold waters also blocked poleward flowing warm currents from the tropics, causing land temperatures to fall to near ice age levels.

The ice sheets appeared to have paused in midstride between 13,000 and 11,500 years ago. That period is called the Younger Dryas, named for an Arctic wildflower, a cold-tolerant plant that grew in Europe. The climate returned to near-glacial conditions midway through the transition from Ice Age to interglacial. Afterward, the warm currents returned. The warming remained permanently, prompting a second episode of melting that led to the present volume of ice by about 6,000 years ago. Following the receding ice sheets, plants and animals began to return to the northern latitudes.

When the ice sheets melted, massive floods raged across the land as water gushed from trapped reservoirs below the glaciers. While flowing under the ice, water surged in vast turbulent sheets that scoured deep grooves in the crust, forming steep ridges carved out of solid bedrock. Each flood continued until the weight of the ice sheet shut off the outlet of the reservoir. When water pressures built up again, another massive surge of meltwater spouted from beneath the glacier and rushed toward the sea. Huge torrents of melt-

Figure 196 *Foraminifera were important limestone builders.*

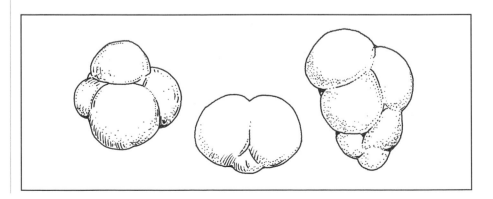

water laden with sediment surged along the Mississippi River toward the Gulf of Mexico, widening the channel to several times its present size. Many other rivers overreached their banks to carve out new floodplains.

The warming paved the way for the Climatic Optimum. This period of unusually warm, wet conditions began 6,000 years ago and lasted for 2,000 years. As the Climatic Optimum unfolded, many regions of the world warmed on average about 5 degrees Celsius. The melting ice caps released massive floodwaters into the sea, raising sea levels 300 feet higher then when the Holocene began.

The inland seas filled with sediments. Subsequent uplifting drove out the waters, leaving behind salt lakes. Great Salt Lake in Utah is today only a remnant of a vast inland sea. During a long wet period between 12,000 and 6,000 years ago, it expanded to several times its current size and flooded the nearby salt flats.

MEGAHERBIVORE EXTINCTION

About 3 million years ago, the uplifting of the Panama Isthmus separating North and South America created an effective barrier to isolate Atlantic and Pacific marine species, which began to go their separate evolutionary ways. Simultaneously, extinctions impoverished the once rich fauna of the western Atlantic. Meanwhile, South America, which had been isolated from the rest of the world for 80 million years, witnessed a vigorous exchange of terrestrial species with North America across the Panama land bridge, causing many species unable to compete with the new arrivals to go extinct.

The connecting landform between North and South America halted the flow of cold water currents from the Atlantic into the Pacific. In addition, the closing of the Arctic Ocean from warm Pacific currents helped initiate the Pleistocene glaciation. Unlike ice ages of the past, the Pleistocene was unusual for its low extinction rates possibly because species were already well adapted to colder conditions. Those species living today are highly robust, having withstood the extreme environmental swings over the last 3 million years, when glaciers spanned much of the Northern Hemisphere.

When the last ice age was drawing to a close between 12,000 and 10,000 years ago, an unusual extinction killed off large terrestrial plant-eating mammals called megaherbivores. These animals grew to enormous size and roamed the ice-free regions in many parts of the Northern Hemisphere. The giantism might have resulted from similar circumstances that led to the great size of some dinosaurs, including an abundant food supply and lack of predation.

Woolly rhinos, mammoths, and Irish elk disappeared in Eurasia. The great buffalo, giant hartebeests, and giant horses disappeared in Africa. More than 80 percent of the large mammals and a significant number of bird species disappeared from Australia. Meanwhile, giant ground sloths (Fig. 197), mastodons, and woolly mammoths (Fig. 198) disappeared from North America. A possible exception was the dwarf woolly mammoth, which might have survived in the Arctic until about 4,000 years ago. The loss of these animals also caused their main predators, the American lion, saber-tooth tiger, and dire wolf, to go extinct.

About 8,000 to 9,000 years ago, conditions were much hotter and dryer than today. The global environment reacted to the changing climate with declining forests and expanding grasslands. The climate change disrupted the food chains of many large animals. When deprived of their food resources, they simply vanished. Also by this time, humans were becoming proficient hunters and roamed northward on the heels of the retreating glaciers. On their journey, they encountered an abundance of wildlife, many species of which they might have hunted to extinction.

In North America, 35 classes of mammals and 10 classes of birds went extinct. The extinctions occurred between 13,000 and 10,000 years ago, with the greatest die out peaking around 11,000 years ago. Most mammals adversely affected were large herbivores weighing more than 100 pounds, with many weighing up to a ton or more. Unlike earlier episodes of mass extinction, this event did not significantly affect small mammals, amphibians, reptiles, and marine invertebrates. Strangely, after having survived several previous

Figure 197 *Giant ground sloths became extinct at the end of the last ice age.*

Figure 198 *The woolly mammoth went extinct at the end of the last ice age.*

periods of glaciation over the past 2 to 3 million years, these large mammals succumbed to extinction following the Ice Age.

During this time, Ice Age peoples occupied many parts of North America. Their spear points have been found resting among the remains of giant mammals, including mammoths, mastodons, tapirs, native horses, and camels. These people crossed into North America from Asia over a land bridge formed by the draining of the Bering Sea and moved through an ice-free corridor east of the Canadian Rockies. Instead of migrating to North America in several waves, however, small bands of nomadic hunters probably crossed the ancient land bridges in pursuit of big game and ended up in the New World purely by accident. The human hunters arriving from Asia sped across the virgin continent, following migrating herds of large herbivores, which they apparently hunted to extinction.

GLACIAL GEOLOGY

Much of the landscape in the northern latitudes owes its unusual geography to massive glaciers that swept down from the polar regions during the last ice age. Glaciers excavated some of most monumental landforms. Outwash streams from glacial meltwater carved out many peculiar landscapes. The power of glacial erosion is well demonstrated by deep-sided valleys carved out of mountain slopes by thick sheets of flowing ice (Fig. 199) a mile or more thick. Glacial erosion radically modified the shapes of stream valleys occupied

Figure 199 *Dome Glacier cascading from the Columbia Ice Fields, showing the moraine-covered tongue of the Athabaska Glacier and abandoned lateral and recessional moraines, Alberta, Canada.*

(Photo by F. O. Jones, courtesy USGS)

by glaciers. The process is most active near the head of a glacier, where the ice deepens and flattens the gradient of the valley. Small hills or knobs in valleys overridden by glaciers are rounded and smoothed by abrasion.

Glaciers flowing down mountain peaks gouged large pits called cirques, from French meaning "circle." They are semicircular basins or indentations with steep walls high on a mountain slope at the head of a valley. The expansion of adjacent cirques by glacial erosion creates arêtes, horns, and cols. An arête, from French meaning "fish bone," is a sharp crested, serrated, or knife-edged ridge that separates the heads of abutting cirques. It also forms a dividing ridge between two parallel valley glaciers. A col is a sharp-edged or saddle-shaped pass in a mountain range formed by the headward erosion where cirques meet or intercept each other. When three or more cirques erode toward a common point, they form a triangular peak called a horn, such as the famous Matterhorn in the Swiss Alps.

The glaciers extended far down the valleys, grinding rocks on the valley floors as the ice advanced and receded. In effect, a river of solid ice embedded with rocks moved along the valley floors, grinding them down like a giant file as the glacier flowed back and forth over them. The advancing glaciers left parallel furrows called glacial striae on the valley floors as they sliced down mountainsides. Miles from existing glaciers are large areas of polished and deeply furrowed rocks, and great heaps of rocks marked the extent of former glaciers. Many of the northern lands are dotted with glacial lakes developed from deep pits excavated by roving glaciers.

Most evidence for extensive glaciation is found in moraines, tillites, and other glacially deposited rocks. Moraines (Fig. 200) are composed of rock and material carried by a glacier and deposited in a regular, linear pattern. They are named according to their position in relation to the glacier. A ground moraine is an irregular carpet of till mainly composed of clay, silt, and sand deposited under a glacier. It is the most prevalent type of continental glacial deposit. A terminal moraine is a ridge of erosional debris deposited by the melting forward margin of a glacier. It is a ridgelike mass of glacial debris formed by the foremost glacial snout and deposited at the outermost edge of glacial advance.

Many parts of the upper midwestern and northeastern United States were eroded down to the granite bedrock. The debris was left in great heaps.

Figure 200 Mount Powell from a remnant of preglacial erosion surface with glacial moraine at the top of Dora Mountain, Summit County, Colorado.

(Photo by O. Twito, courtesy USGS)

Figure 201 *A drumlin in Middlesex County, Massachusetts.*

(Photo by J. C. Russell, courtesy USGS)

The glacially derived sediments covered much of the landscape, burying older rocks under thick layers of till. Glacial till is nonstratified or mixed material comprising clay and boulders deposited directly by glacial ice. Basal till at the base of a glacier was usually laid down under it. Ablation till on or near the surface of a glacier was deposited when the ice melted. Tillites are a consolidated mixture of boulders and pebbles in a clay matrix. They were deposited by glacial ice and are known to exist on every continent.

In some areas, older deposits were buried under thick deposits of glacial till, forming elongated hillocks aligned in the same direction called drumlins (Fig. 201). Drumlins are tall and narrow at the upstream end of the glacier and slope to a low, broad tail. The hills appear in concentrated fields in North America, Scandinavia, Britain, and other areas once covered by ice. Drumlin fields might contain as many as 10,000 knolls, looking like rows of eggs lying on their sides. Drumlins are the least understood of all glacial landforms. How they attained their characteristic oval shape remains a mystery. Perhaps the extensive drumlin fields of North America were created during cataclysmic flood processes during melting of the vast ice sheets.

Similar to a drumlin is roche moutonnée, from French meaning "fleecy rock." The term was applied to glaciated outcrops because they resemble the backs of sheep, thus prompting the name sheepback rock. Roche moutonnée is a glaciated bedrock surface with asymmetrical mounds of varying shapes. The up-glacier side has been glacially scoured and smoothly abraded. The down-glacier side has steeper, jagged slopes, resulting from glacial plucking.

The ridges dividing the two sides of a roche moutonnée are perpendicular to the general flow of the ice sheets.

At the margins of the ice sheets are rugged periglacial regions. These features developed along the tip of the ice and were directly controlled by the glacier. Cold winds blowing off the ice sheets affected the climate of the glacial margins and helped create periglacial conditions. The zone was dominated by such processes as frost heaving, frost splitting, and sorting, which created immense boulder fields out of what was once solid bedrock.

Erratic boulders (Fig. 202) are glacially transported rocks embedded in glacial till or exposed on the surface. They indicate the direction of glacial flow. They range in size from pebbles to massive boulders and have traveled as far as 500 miles or more. Erratics are composed of distinctive rock types that can be traced to their places of origin. Indicator boulders are erratics of known origin used to locate the source area and the distance traveled for any given glacial till. Their identifying features include a distinctive appearance, unique mineral assemblage, or characteristic fossil content. The erratics are often arranged in a boulder train, forming a line or series of rocks originating from the same bedrock source and extending in the direction of glacial movement. A boulder fan is a conically shaped deposit containing distinctive erratics derived from an outcrop at the apex of the fan.

Glacial drift includes all rock material deposited by glaciers or glacier-fed streams and lakes, where the greatest thicknesses are attained in buried valleys.

Figure 202 *A glacial boulder field, Tulare County, California.*

(Photo by F. E. Matthes, courtesy USGS)

Figure 203 *An esker in Wisconsin.*

(Photo by W. C. Alden, courtesy USGS)

Drift is divided into two types of material. One is till deposited directly by glacial ice and shows little or no sorting or stratification (layering) as though dumped haphazardly. The other is stratified drift, which is a well-sorted, layered material transported and deposited by glacial meltwater. Streams flowing from melting glaciers rework part of the glacial material, some of which is carried into standing bodies of water, forming banded deposits called glacial varves.

Long, sinuous sand deposits called eskers (Fig. 203) formed out of glacial debris from outwash streams. They are winding, steep-walled ridges that can extend up to 500 miles in length but seldom exceed more than 1,000 to 2,000 feet wide and 150 feet high. Eskers appear to have been created by streams running through tunnels beneath the ice sheet. When the ice melted, the old stream deposits were left standing as a ridge. Eskers appear to have been deposited in channels beneath or within slow-moving or stagnant glacial ice. Their general orientation runs at right angles to the glacial edge. At the margin of a glacial lake, they might form river deltas.

Kames are mounds composed chiefly of stratified sand and gravel formed at or near the snout of an ice sheet or deposited at the margin of a melting glacier. They are found in areas where large quantities of coarse material are available during the slow melting of glacial ice. Meltwater is present in sufficient quantities to redistribute the debris and deposit the sediments at the margins of the decaying ice mass. Most kames are low, irregularly conical mounds of roughly layered glacial sand and gravel that often occur in clusters. They accompany the terminal moraine region of valley and continental glaciers. They appear to represent sediment fillings of openings in stagnant ice. Many kames probably formed when streams flowed off the tops of glaciers onto the bare ground, dropping sediment into thick piles.

The Ice Age is still with us, only we are fortunate to live in a warmer period between glaciations. Perhaps in another couple thousand years, massive ice sheets will again be on the rampage, wiping out everything in their paths. Rubble from demolished northern cities would be bulldozed hundreds of miles to the south, gripped in the frozen jaws of the advancing ice. Global temperatures would plummet as plants and animals, including humans, scramble for the warmth of the tropics.

CONCLUSION

After taking a journey through geologic time, we can better appreciate the wide variety of life on Earth today. However, as shown here, species have been struck down several times in Earth history by extinction.

Life then bounded back with greater diversity, so that today's world is filled with a vast array of species far more abundant than has ever existed before. A tremendous shame would be upon us if we should destroy nature's bounty by despoiling the habitats of all creatures great and small.

GLOSSARY

abyss   the deep ocean, generally over a mile in depth

Acanthostega (ah-KAN-the-stay-ga)   an extinct primitive Paleozoic amphibian

accretion   the accumulation of celestial gas and dust by gravitational attraction into a planetesimal, asteroid, moon, or planet

age   a geologic time interval smaller than an epoch

albedo   the amount of sunlight reflected from an object and dependent on the object's color and texture

alluvium (ah-LUE-vee-um)   stream-deposited sediment

alpine glacier   a mountain glacier or a glacier in a mountain valley

amber   fossil tree resin that has achieved a stable state after ground burial via chemical change and the loss of volatile constituents

ammonite (AM-on-ite)   a Mesozoic cephalopod with flat, spiral shells

amphibian   a cold-blooded, four-footed vertebrate belonging to a class midway in evolutionary development; fish and reptiles

andesite   a volcanic rock intermediate between basalt and rhyolite

angiosperm (AN-jee-eh-sperm)   a flowering plant that reproduces sexually with seeds

annelid (A-nil-ed)   a wormlike invertebrate characterized by a segmented body with a distinct head and appendage

archaeocyathan (AR-key-ah-sy-a-than)   an ancient Precambrian organism resembling sponges and coral; it built early limestone reefs

277

Archaeopteryx (AR-key-op-the-riks)   a primitive Jurassic crow-sized bird characterized with teeth and a bony tail

archea (AR-key-ah)   a primitive bacteria-like organism living in high-temperature environments

Archean (AR-key-an)   a major eon of the Precambrian from 4.0 to 2.5 billion years ago

arthropod (AR-threh-pod)   the largest group of invertebrates, including crustaceans and insects, characterized by segmented bodies, jointed appendages, and exoskeletons

asteroid   a rocky or metallic body, orbiting the Sun between Mars and Jupiter, and leftover from the formation of the solar system

asteroid belt   a band of asteroids orbiting the Sun between the orbits of Mars and Jupiter

Azoic eon   a term applied to the first half-billion years of Earth's history

Baltica (BAL-tik-ah)   an ancient Paleozoic continent of Europe

barrier island   a low, elongated coastal island that parallels the shoreline and protects the beach from storms

basalt (bah-SALT)   a dark volcanic rock rich in iron and magnesium and usually quite fluid in the molten state

batholith (BA-tha-lith)   the largest of intrusive igneous bodies and more than 40 square miles on its uppermost surface

bedrock   solid layers of rock lying beneath younger material

belemnite (BEL-em-nite)   an extinct Mesozoic cephalopod with a bullet-shaped internal shell

bicarbonate   an ion created by the action of carbonic acid on surface rocks; marine organisms use the bicarbonate along with calcium to build supporting structures composed of calcium carbonate

big bang   a theory for the creation of the universe, dealing with the initiation of all matter

biogenic   sediments composed of the remains of plant and animal life such as shells

biomass   the total mass of living organisms within a specific habitat

biosphere   the living portion of Earth that interacts with all other biologic and geologic processes

bivalve   a mollusk with a shell comprising two hinged valves, including oysters, muscles, and clams

black smoker   superheated hydrothermal water rising to the surface at a midocean ridge; the water is supersaturated with metals, and when exiting through the seafloor, the water quickly cools and the dissolved metals precipitate, resulting in black, smokelike effluent

blastoid   an extinct Paleozoic echinoderm similar to a crinoid with a body resembling a rosebud

brachiopod (BRAY-key-eh-pod) a marine, shallow-water invertebrate with bivalve shells similar to mollusks and plentiful in the Paleozoic

bryophyte (BRY-eh-fite) nonflowering plants comprising mosses, liverworts, and hornworts

bryozoan (BRY-eh-zoe-an) a marine invertebrate that grows in colonies and is characterized by a branching or fanlike structure

calcite a mineral composed of calcium carbonate

caldera (kal-DER-eh) a large, pitlike depression at the summits of some volcanoes and formed by great explosive activity and collapse

calving formation of icebergs by glaciers breaking off upon entering the ocean

Cambrian explosion a rapid radiation of species that occurred as a result of a large adaptive space, including numerous habitats and mild climate

carbonaceous (KAR-beh-NAY-shes) a substance containing carbon, namely sedimentary rocks such as limestone and certain types of meteorites

carbonate a mineral containing calcium carbonate such as limestone and dolostone

carbon cycle the flow of carbon into the atmosphere and ocean, the conversion to carbonate rock, and the return to the atmosphere by volcanoes

Cenozoic (SIN-eh-zoe-ik) an era of geologic time comprising the last 65 million years

cephalopod (SE-feh-lah-pod) marine mollusks including squids, cuttlefish, and octopuses that travel by expelling jets of water

chalk a soft form of limestone composed chiefly of calcite shells of microorganisms

chert an extremely hard cryptocrystalline quartz rock resembling flint

chondrule (KON-drule) rounded granules of olivine and pyroxine found in stony meteorites called chondrites

cirque a glacial erosional feature, producing an amphitheater-like head of a glacial valley

class in systematics, the category of plants and animals below a phylum comprising several orders

coal a fossil fuel deposit originating from metamorphosed plant material

coelacanth (SEE-leh-kanth) a lobe-finned fish originating in the Paleozoic and presently living in deep seas

coelenterate (si-LEN-the-rate) multicellular marine organisms, including jellyfish and corals

comet a celestial body believed to originate from a cloud of comets that surrounds the Sun and develops a long tail of gas and dust particles when traveling near the inner solar system

conglomerate (kon-GLOM-er-ate) a sedimentary rock composed of welded fine-grained and coarse-grained rock fragments

conodont a Paleozoic toothlike fossil probably from an extinct marine vertebrate

continent a landmass composed of light, granitic rock that rides on denser rocks of the upper mantle

continental drift the concept that the continents drift across the surface of Earth

continental glacier an ice sheet covering a portion of a continent

continental shelf the offshore area of a continent in a shallow sea

continental slope the transition from the continental shelf to the deep-sea basin

convection a circular, vertical flow of a fluid medium by heating from below; as materials are heated, they become less dense and rise, cool down, become more dense, and sink

coquina (koh-KEY-nah) a limestone comprised mostly of broken pieces of marine fossils

coral a large group of shallow-water, bottom-dwelling marine invertebrates comprising reef-building colonies common in warm waters

Cordillera (kor-dil-ER-ah) a range of mountains that includes the Rockies, Cascades, and Sierra Nevada in North America and the Andes in South America

correlation (KOR-el-LAY-shen) the tracing of equivalent rock exposures over distance usually with the aid of fossils

craton (CRAY-ton) the ancient, stable interior region of a continent, usually composed of Precambrian rocks

crinoid (KRY-noid) an echinoderm with a flowerlike body atop a long stalk of calcite disks

crossopterygian (CROS-op-tary-gee-an) an extinct Paleozoic fish thought to give rise to terrestrial vertebrates

crust the outer layers of a planet's or a moon's rocks

crustacean (kres-TAY-shen) an arthropod characterized by two pairs of antenna-like appendages forward of the mouth and three pairs behind it, including shrimps, crabs, and lobsters

diapir (DIE-ah-per) the buoyant rise of a molten rock through heavier rock

diatom (DIE-ah-tom) microplants whose fossil shells form siliceous sediments called diatomaceous earth

dinoflagellate (DIE-no-FLA-jeh-late) planktonic single-celled organisms important in marine food chains

dolomite (DOH-leh-mite) a sedimentary rock formed when calcium in limestone is replaced by magnesium

drumlin a hill of glacial debris facing in the direction of glacial movement

dune a ridge of windblown sediments usually in motion

earthquake   the sudden rupture of rocks along active faults in response to geologic forces within Earth

East Pacific Rise   a midocean ridge spreading system running north–south along the eastern side of the Pacific; the predominant location where hot springs and black smokers were discovered

echinoderm (I-KY-neh-derm)   marine invertebrates, including starfish, sea urchins, and sea cucumbers

echinoid (I-KY-noid)   a group of echinoderms including sea urchins and sand dollars

ecliptic   the plane of Earth's orbit around the Sun

ecology   the interrelationships between organisms and their environment

ecosystem   a community of organisms and their environment functioning as a complete, self-contained biological unit

Ediacaran   a group of unique, extinct, late-Precambrian organisms

environment   the complex physical and biological factors that act on an organism to determine its survival and evolution

eon   the longest unit of geologic time, roughly about 1 billion years or more in duration

epoch   a geologic time unit shorter than a period and longer than an age

era   a unit of geologic time below an eon, consisting of several periods

erosion   the wearing away of surface materials by natural agents such as wind and water

erratic   a glacially deposited boulder far from its source

esker   a long narrow ridge of sand and gravel from a glacial outwash stream

eukaryote (yu-KAR-ee-ote)   a highly developed organism with a nucleus that divides genetic material in a systematic manner

eurypterid (yu-RIP-te-rid)   a large Paleozoic arthropod related to the horseshoe crab

evaporite   the deposition of salt, anhydrite, and gypsum from evaporation in an enclosed basin of stranded seawater

evolution   the tendency of physical and biological factors to change with time

exoskeleton   the hard outer protective covering of invertebrates including cuticles and shells

extinction   the loss of large numbers of species over a short duration, sometimes marking the boundaries of geologic periods

extrusive (ik-STRU-siv)   an igneous volcanic rock ejected onto Earth's surface

family   in systematics, the category of plants and animals under order and comprising several genera

fault   a break in crustal rocks caused by Earth movements

feldspar   a group of rock-forming minerals comprising about 60 percent of Earth's crust and an essential component of igneous, metamorphic, and sedimentary rock

fissure   a large crack in the crust through which magma might escape from a volcano

fluvial (FLUE-vee-al)   stream-deposited sediment

foraminifer (FOR-eh-MI-neh-fer)   a calcium carbonate-secreting organism that lives in the surface waters of the oceans; after death, their shells form the primary constituents of limestone and sediments deposited onto the seafloor

formation   a combination of rock units that can be traced over distance

fossil   any remains, impression, or traces in rock of a plant or animal of a previous geologic age

fossil fuel   an energy source derived from ancient plant and animal life that includes coal, oil, and natural gas; when ignited, these fuels release carbon dioxide that was stored in Earth's crust for millions of years

fusulinid (FEW-zeh-LIE-nid)   a group of extinct foraminiferans resembling a grain of wheat

galaxy   a large, gravitationally bound cluster of stars

gastrolith (GAS-tra-lith)   a stone ingested by an animal used to grind food

gastropod (GAS-tra-pod)   a large class of mollusks, including slugs and snails, characterized by a body protected by a single shell that is often coiled

genus   in systematics, the category of plants and animals under family and comprising several species

geologic column   the total thickness of geologic units in a region

geothermal   the generation of hot water or steam by hot rocks in Earth's interior

geyser (GUY-sir)   a spring that ejects intermittent jets of steam and hot water

glacier   a thick mass of moving ice occurring where winter snowfall exceeds summer melting

Glossopteris (GLOS-opt-ter-is)   a late Paleozoic plant that existed on the southern continents but not on the northern continents, thereby confirming the existence of Gondwana

gneiss (nise)   a foliated metamorphic rock with a similar composition as granite

Gondwana (gone-DWAN-ah)   a southern supercontinent of Paleozoic time, comprised of Africa, South America, India, Australia, and Antarctica; it broke up into the present continents during the Mesozoic era

granite   a coarse-grained, silica-rich igneous rock consisting primarily of quartz and feldspars

graptolite (GRAP-the-lite)   extinct Paleozoic planktonic animals resembling tiny stems

greenhouse effect   the trapping of heat in the lower atmosphere principally by water vapor and carbon dioxide

greenstone   a green, weakly formed, metamorphic igneous rock

greenstone belt   a mass of Archean metamorphosed igneous rock

gypsum   a white, colorless calcium sulfate mineral formed during the evaporation of brine pools; when calcined, it forms plaster of Paris

Hallucigenia (HA-loose-ah-gen-ia)   an unusual animal of the early Cambrian with stiltlike legs and multiple mouths along the back

hexacoral   coral with six-sided skeletal walls

horn   a peak on a mountain formed by glacial erosion

hydrocarbon   a molecule consisting of carbon chains with attached hydrogen atoms

Iapetus Sea (EYE-ap-I-tus)   a former sea that occupied a similar area as the present Atlantic Ocean prior to the assemblage of Pangaea

ice age   a period of time when large areas of Earth were covered by massive glaciers

ichthyosaur (IK-the-eh-sore)   an extinct Mesozoic aquatic reptile with a streamlined body and long snout

Ichthyostega (IK-the-eh-ste-ga)   an extinct primitive Paleozoic fishlike amphibian

impact   the point on the surface upon which a celestial object lands

index fossil   a representative fossil that identifies the rock strata in which it is found

interglacial   a warming period between glacial periods

invertebrate   an animal with an external skeleton such as shellfish and insects

iridium   a rare isotope of platinum, relatively abundant on meteorites

island arc   volcanoes landward of a subduction zone parallel to a trench and above the melting zone of a subducting plane

karst   a terrain comprised of numerous sinkholes in limestone

lacustrine (leh-KES-trene)   inhabiting or produced in lakes

Laurasia (lure-AY-zha)   a northern supercontinent of Paleozoic time, consisting of North America, Europe, and Asia

Laurentia (lure-IN-tia)   an ancient North American continent

lava   molten magma that flows out onto the surface

limestone   a sedimentary rock composed of calcium carbonate secreted from seawater by invertebrates and whose skeletons compose the bulk of deposits

lithosphere (LI-tha-sfir)   the rocky outer layer of the mantle that includes the terrestrial and oceanic crusts; the lithosphere circulates between Earth's surface and mantle by convection currents

lithospheric   a segment of the lithosphere, the upper layer plate of the mantle, involved in the interaction of other plates in tectonic activity

loess   a thick deposit of airborne dust

lungfish   a bony fish that breathes on land and in water

lycopod (LIE-keh-pod)   the first ancient trees of Paleozoic forests; today comprising club mosses and liverworts

Lystrosaurus   an ancient, extinct, mammal-like reptile with large down-turned tusks

magma   a molten rock material generated within Earth and the constituent of igneous rocks

mantle   the part of a planet below the crust and above the core, composed of dense rocks that might be in convective flow

maria   dark plains on the lunar surface produced by massive basalt floods

marsupial (mar-SUE-pee-al)   a primitive mammal that weans underdeveloped infants in a belly pouch

megaherbivore   a large, plant-eating animal such as an elephant or extinct mastodon

Mesozoic (meh-zeh-ZOE-ik)   literally the period of middle life, referring to a period between 250 and 65 million years ago

metamorphic rock   a rock crystallized from previous igneous, metamorphic, or sedimentary rocks created under conditions of intense temperatures and pressures without melting

metamorphism (me-teh-MORE-fi-zem)   recrystallization of previous igneous, metamorphic, and sedimentary rocks under extreme temperatures and pressures without melting

metazoan   primitive multicellular animal with cells differentiated for specific functions

meteorite   a metallic or stony celestial body that enters Earth's atmosphere and impacts on the surface

methane   a hydrocarbon gas liberated by decomposing organic matter and a major constituent of natural gas

microfossil   a fossil that must be studied with a microscope; used for dating drill cuttings

mollusk (MAH-lusk)   a large group of invertebrates, including snails, clams, squids, and extinct ammonites, characterized by an internal and external shell surrounding the body

monotreme   egg-laying mammals including the platypus and echidna

moraine   a ridge of erosional debris deposited by the melting margin of a glacier

nautiloid (NOT-eh-loid)   shell-bearing cephalopods abundant in the Paleozoic with only the nautilus surviving

nebula   an extended astronomical object with a cloudlike appearance; some nebulae are galaxies, others are clouds of dust and gas within our galaxy

Neogene   the Cenozoic Miocene and Pliocene epochs

nutrient   a food substance that nourishes living organisms

oolite (OH-eh-lite)   small, rounded grains in limestone

Oort cloud   a collection of comets that surround the Sun about a light-year away

ophiolite (OH-fi-ah-lite)   oceanic crust thrust upon continents by plate tectonics

ore body   the accumulation of metal-bearing ores where the hot, hydrothermal water moving upward toward the surface mixes with cold seawater penetrating downward

orogens (ORE-ah-gins)   eroded roots of ancient mountain ranges

orogeny (oh-RAH-ja-nee)   an episode of mountain building by tectonic activity

outgassing   the loss of gas from within a planet as opposed to degassing, the loss of gas from meteorites

ozone   a molecule consisting of three atoms of oxygen in the upper atmosphere and that filters out harmful ultraviolet radiation from the Sun

Paleogene   the Cenozoic Paleocene, Eocene, and Oligocene epochs

paleomagnetism   the study of Earth's magnetic field, including the position and polarity of the poles in the past

paleontology (pay-lee-ON-TAL-o-gee)   the study of ancient life-forms based on the fossil record of plants and animals

Paleozoic (PAY-lee-eh-ZOE-ic)   the period of ancient life, between 570 and 250 million years ago

Pangaea (pan-GEE-a)   a Paleozoic supercontinent that included all the lands of Earth

Panthalassa (PAN-the-lass-ah)   the global ocean that surrounded Pangaea

peridotite   the most common ultramafic rock type in the mantle

period   a division of geologic time longer than an epoch and included in an era

photosynthesis   the process by which plants form carbohydrates from carbon dioxide, water, and sunlight

phyla   groups of organisms that share similar body forms

phytoplankton   marine or freshwater microscopic, single-celled, freely drifting plant life

pillow lava   lava extruded on the ocean floor giving rise to tabular shapes

placer (PLAY-ser)   a deposit of rocks left behind by a melting glacier; any ore deposit that is enriched by stream action

placoderm   an extinct class of chordates, fish with armorlike plates and articulated jaws

planetesimals   small celestial bodies that accreted during the early stages of the solar system

plate tectonics   the theory that accounts for the major features of Earth's surface in terms of the interaction of lithospheric plates

pluton (PLUE-ton)   an underground body of igneous rock younger than the rocks that surround it; it is formed where molten rock assimilates older rocks

prebiotic   conditions on the early Earth prior to the introduction of life processes

precipitation   the deposition of minerals from seawater

primary producer   the lowest member of a food chain

primordial   pertaining to the primitive conditions that existed during early stages of development

prokaryote (pro-KAR-ee-ote)   a primitive organism that lacks a nucleus

protist (PRO-tist)   a unicellular organism, including bacteria, protozoans, algae, and fungi

pseudofossil   a fossil-like body such as a concretion

pterosaur (TER-eh-sore)   extinct Mesozoic flying reptiles with batlike wings

radiolarian   a microorganism with shells made of silica comprising a large component of siliceous sediments

radiometric dating   the age determination of an object by radiometric and chemical analysis of stable versus unstable radioactive elements

redbed   a sedimentary rock cemented with iron oxide

reef   the biological community that lives at the edge of an island or continent; the shells from dead organisms form a limestone deposit

regression   a fall in a sea level, exposing continental shelves to erosion

reptile   an air-breathing, cold-blooded animal that lays eggs and is usually covered with scales

Rodinia   a Precambrian supercontinent whose breakup sparked the Cambrian explosion of species

sandstone   a sedimentary rock consisting of sand grains cemented together

sedimentary rock   a rock composed of fragments cemented together

shale   a fine-grained fissile sedimentary rock of consolidated mud or clay

shield   areas of the exposed Precambrian nucleus of a continent

species   groups of organisms that share similar characteristics and are able to breed among themselves

spherules   small, spherical, glassy grains found on certain types of meteorites, on lunar soils, and at large meteorite impact sites

strata   layered rock formations; also called beds

stromatolite (STRO-mat-eh-lite)   a calcareous structure built by successive layers of bacteria or algae and that have existed for the past 3.5 billion years

subduction zone   a region where an oceanic plate dives below a continental plate into the mantle; ocean trenches are the surface expression of a subduction zone

supernova   an enormous stellar explosion in which all but the inner core of a star is blown off into interstellar space

tectonic activity   the formation of Earth's crust by large-scale movements throughout geologic time

tephra (TEH-fra)   solid material ejected into the air by a volcanic eruption

terrestrial   all phenomena pertaining to Earth

Tethys Sea (THE-this)   the hypothetical midlatitude region of the oceans separating the northern and southern continents of Laurasia and Gondwana several hundred million years ago

tetrapod   a four-footed vertebrate

thecodont (THEE-keh-daunt)   an ancient primitive reptile that gave rise to dinosaurs, crocodiles, and birds

therapsid (the-RAP-sid)   ancient reptile ancestors of the mammals

therian   animals that have live births such as mammals

thermophilic   relating to primitive organisms that live in hot-water environments

tide   a bulge in the ocean produced by the Sun's and Moon's gravitational forces on Earth's oceans; the rotation of Earth beneath this bulge causes the rising and lowering of the sea level

till   nonstratified material deposited directly by glacial ice as it recedes and consolidated into tillite

tillite   a sedimentary deposited composed of glacial till

transgression   a rise in sea level that causes flooding of the shallow edges of continental margins

trilobite (TRY-leh-bite)   an extinct marine arthropod, characterized by a body divided into three lobes, each bearing a pair of jointed appendages, and a chitinous exoskeleton

tundra   permanently frozen ground at high latitudes of the Arctic regions

type section   a sequence of strata that was originally described as constituting a stratigraphic unit and serves as a standard of comparison for identifying similar widely separated units

ultraviolet   the invisible light with a wavelength shorter than visible light and longer than X rays

varves   thinly laminated lake bed sediments deposited by glacial meltwater

vertebrates   animals with an internal skeleton, including fish, amphibians, reptiles, and mammals

volcanism   any type of volcanic activity

volcano   a fissure or vent in the crust through which molten rock rises to the surface to form a mountain

zooplankton   small, free-floating or poorly swimming marine or freshwater animal life

BIBLIOGRAPHY

PLANET EARTH

Allegre, Claud J., and Stephen H. Snider. "The Evolution of the Earth." *Scientific American* 271 (October 1994): 66–75.

Black, David C. "Worlds Around Other Stars." *Scientific American* 264 (January 1991): 76–82.

Hartley, Karen. "A New Window on Star Birth." *Astronomy* 17 (March 1989): 32–36.

Horgan, John. "In the Beginning." *Scientific American* 264 (February 1991): 117–125.

Kasting, James F., Owen B. Toon, and James B. Pollack. "How Climate Evolved on the Terrestrial Planets." *Scientific American* 258 (February 1988): 90–98.

Kerr, Richard A. "Making the Moon, Remaking the Earth." *Science* 243 (March 17, 1989): 1433–1435.

Monastersky, Richard. "The Rise of Life on Earth." *National Geographic* 194 (March 1998): 54–81.

Newsom, Horton E., and Kenneth W. W. Sims. "Core Formation During Early Accretion of the Earth." *Science* 252 (May 17, 1991): 926–933.

Orgel, Leslie E. "The Origin of Life on the Earth." *Scientific American* 271 (October 1994): 77–83.

Simpson, Sarah. "Life's First Scalding Steps." *Science News* 155 (January 9, 1999): 24–26.

Taylor, S. Ross, and Scott M. McLennan. "The Evolution of Continental Crust." *Scientific American* 274 (January 1996): 76–81.

Waldrop, Mitchell M. "Goodbye to the Warm Little Pond?" *Science* 250 (November 23, 1990): 1078–1080.

ARCHEAN ALGAE

Dickinson, William R. "Making Composite Continents." *Nature* 364 (July 22, 1993): 284–285.

Duve, Christian. "The Birth of Complex Cells." *Scientific American* 274 (April 1996): 50–57.

Horgon, John. "Off to an Early Start." *Scientific American* 269 (August 1993): 24.

Jones, Richard C., and Anthony N. Stranges. "Unraveling Origins, the Archean." *Earth Science* 42 (Winter 1989): 20–22.

Lowe, Donald R., et al. "Geological and Geochemical Record of 3400-Million-Year-Old Terrestrial Meteorite Impacts." *Science* 245 (September 1, 1989): 959–962.

Moores, Eldridge. "The Story of Earth." *Earth* 5 (December 1996): 30–33.

Pool, Robert. "Pushing the Envelope of Life." *Science* 247 (January 12, 1990): 158–160.

Preiser, Rachel. "The First Land." *Discover* 16 (December 1995): 32–33.

Radetsky, Peter. "Life's Crucible." *Earth* 6 (February 1998): 34–41.

Taylor, Stuart Ross. "Young Earth like Venus?" *Nature* 350 (April 4, 1991): 376–377.

Weiss, Peter. "Land Before Time." *Earth* 8 (February 1998): 29–33.

York, Derek. "The Earliest History of the Earth." *Scientific American* 268 (January 1993): 90–96.

Zimmer, Carl. "Ancient Continent Opens Window on the Early Earth." *Science* 286 (December 17, 1999): 2254–2256.

PROTEROZOIC METAZOANS

Edgar, Blake. "The Ultimate Missing Link." *Earth* 6 (October 1997): 30–31.

Gould, Stephen Jay. "The Evolution of Life on the Earth." *Scientific American* 271 (October 1994): 85–91.

Hoffman, Paul F., and David P. Schrag. "Snowball Earth." *Scientific American* 282 (January 2000): 68–75.

Kerr, Richard A. "Pushing Back the Origins of Animals." *Science* 279 (February 6, 1998): 803–804.

Knoll, Andrew H. "End of the Proterozoic Eon." *Scientific American* 265 (October 1991): 64–73.

Kunzig, Robert. "Birth of a Nation." *Discover* 11 (February 1990): 26–27.

Monastersky, Richard. "Life Grows Up." *National Geographic* 194 (April 1998): 100–115.

———. "Jump-Start for the Vertebrates." *Science News* 149 (February 3, 1996): 74–75.

———. "Oxygen Upheaval." *Science News* 142 (December 12, 1992): 412–413.

Shurkin, Joel, and Tom Yulsman. "Assembling Asia." *Earth* 4 (June 1995): 52–59.

Wright, Karen. "When Life Was Odd." *Discover* 18 (March 1997): 52–61.

Zimmer, Carl. "An Explosion Defused." *Discover* 17 (December 1996): 52.

CAMBRIAN INVERTEBRATES

Beardsley, Tim. "Weird Wonders." *Scientific American* 266 (June 1992): 30–34.

Kerr, Richard A. "Evolution's Big Bang Gets Even More Explosive." *Science* 261 (September 3, 1993): 1274–1275.

———. "The Earliest Mass Extinction?" *Science* 257 (July 31, 1992): 612.

Landman, Neil H. "Luck of the Draw." *Natural History* 100 (December 1991): 68–71.

Levinton, Jeffrey S. "The Big Bang of Animal Evolution." *Scientific American* 267 (November 1992): 84–91.

Mestel, Rosie. "Kimberella's Slippers." *Earth* 6 (October 1997): 24–29.

Monastersky, Richard. "The Edicaran Enigma." *Science News* 148 (July 9, 1995): 28–30.

Morell, Virginia. "Sizing Up Evolutionary Radiations." *Science* 274 (November 29, 1996): 1462–1463.

Morris, S. Conway. "Burgess Shale Faunas and the Cambrian Explosion." *Science* 246 (October 20, 1989): 339–345.

Palmer, Allison, R. "A Fondness for 'Beetles of the Paleozoic'" *Science* 290 (October 6, 2000): 59.

Palmer, Douglas. "Trilobite Fossil Shows Its Muscle." *New Scientist* (November 6, 1993): 20.

Stolzenburg, William. "When Life Got Hard." *Science News* 138 (August 25, 1990): 120–123.

ORDOVICIAN VERTEBRATES

Bower, Bruce. "Fossils Flesh Out Early Vertebrates." *Science News* 133 (January 9, 1988): 21.

Forey, Peter, and Philippe Janvier. "Agnathans and the Origin of Jawed Vertebrates." *Nature* 361 (January 14, 1993): 129–133.

Irion, Robert. "Parsing the Trilobites's Rise and Fall." *Science* 280 (June 19, 1998): 1837.

Kerr, Richard A. "Missing Chunk of North America Found in Argentina." *Science* 270 (December 8, 1995): 1567–1568.

Knox, C. "New View Surfaces of Ancient Atlantic." *Science News* 134 (October 8, 1988): 230.

Monastersky, Richard. "Jump-Start for the Vertebrates." *Science News* 149 (February 3, 1996): 74–75.

Travis, John. "Eye-Opening Gene." *Science News* 151 (May 10, 1997): 288–289.

Rigbey, Sue. "Graptolites Come to Life." *Nature* 362 (March 18, 1993): 209–210.

Svitil, Kathy A. "It's Alive, and It's a Graptolite." *Discover* 14 (July 1993): 18–19.

Weisburd, Stefi. "Facing Up To a Backward Fossil." *Science News* 132 (July 18, 1987): 47.

Zimmer, Carl. "Location, Location, Location." *Discover* 14 (December 1994): 20–32.

SILURIAN PLANTS

Dickinson, William R. "Making Composite Continents." *Nature* 364 (July 22, 1993): 284–285.

Jeffrey, David. "Fossils: Animals of Life Written in Rock." *National Geographic* 168 (August 1985): 182–191.

Kerr, Richard A. "A Half-Billion-Year Head Start For Life on Land." *Science* 258 (November 13, 1992): 1082–1083.

Lewin, Roger. "On the Origin of Insect Wings." *Science* 230 (October 25, 1985): 428–429.

Monastersky, Richard. "Fossil Soil Has the Dirt on Early Microbes." *Science News* 153 (March 7, 1998): 151.

———. "Fossils Push Back Origin of Land Animals." *Science News* 138 (November 10, 1990): 292.

———. "Supersoil." *Science News* 136 (December 9, 1989): 376–377.

Pestrong, Ray. "It's About Time." *Earth Science* 42 (Summer 1989): 14–15.

Preiser, Rachel. "Soft Silurians." *Discover* 16 (November 1996): 36.

Schreeve, James. "Are Algae—Not Coral—Reefs' Master Builders?" *Science* 271 (February 2, 1996): 597–598.

DEVONIAN FISH

Barnes-Svarney, Patricia. "In Search of Ancient Shores." *Earth Science* 40 (Spring 1987): 22.

Dalziel, Ian W. D. "Earth Before Pangaea." *Scientific American* 272 (January 1995): 58–63.

Eastman, Joseph T., and Arthur L. DeVries. "Antarctic Fishes." *Scientific American* 255 (November 1986): 106–114.

Forey, Peter, and Philippe Janvier. "Agnathans and the Origin of Jawed Vertebrates." *Nature* 361 (January 14, 1993): 129–133.

Glausiusz, Josie. "The Old Fish of the Sea." *Discover* 23 (January 1999): 49.

Kerr, Richard A. "Another Impact Extinction?" *Science* 256 (May 1992): 1280.

———. "How Long Does It Take to Build a Moutain?" *Science* 240 (June 24, 1988): 1735.

Landman, Neil H. "Luck of the Draw." *Natural History* 100 (December 1991): 68–71.

Richardson, Joyce R. "Brachiopods." *Scientific American* 255 (September 1986): 100–106.

Roush, Wade. "Living Fossil Fish Is Dethroned." *Science* 277 (September 5, 1997): 1436.

Vogel, Shawna. "Face-To-Face with a Living Fossil." *Discover* 9 (March 1988): 56–57.

Zimmer, Carl. "Breathe Before You Bite." *Discover* 17 (March 1996): 34.

CARBONIFEROUS AMPHIBIANS

Arenz, Jim. "A River Ran Through It." *Earth* 5 (October 1996): 14.

Fulkerson, William, et al. "Energy from Fossil Fuels." *Scientific American* 263 (September 1990): 129–135.

Hannibal, Joseph T. "Quarries Yield Rare Paleozoic Fossils." *Geotimes* 33 (July 1988): 10–13.

Monastersky, Richard. "Out of the Swamps." *Science News* 155 (May 22, 1999): 328–330.

———. "Swamped by Climate Change?" *Science News* 138 (September 22, 1990): 184–186.

———. "Ancient Amphibians Found in Iowa." *Science News* 133 (June 25, 1988): 406.

O'Hanlon, Larry. "Age of the Giant Insects." *Earth* 4 (October 1995): 10.

Rigby, Sue. "Graptolites Come to Life." *Nature* 362 (March 18, 1993): 209–210.

Svitil, Kathy A. "Landlubber Genes." *Discover* 20 (May 2000): 15.

Waters, Tom. "Greetings from Pangaea." *Discover* 13 (February 1992): 38–43.

Weisburd, Stefi. "Forests Made the World Frigid?" *Science News* 131 (January 3, 1987): 9.

Westenberg, Kerri. "From Fins to Feet." *National Geographic* 195 (May 1999): 114–127.

Zimmer, Carl. "Coming onto the Land." *Discover* 16 (June 1995): 120–127.

PERMIAN REPTILES

Berman, David S., et al. "Early Permian Bipedal Reptile." *Science* 290 (November 3, 2000): 969–972.

Crowley, Thomas J., and Gerald R. North. "Abrupt Climate Change and Extinction Events in Earth History." *Science* 240 (May 20, 1988): 996–1001.

Erwin, Douglas H. "The Mother of Mass Extinctions." *Scientific American* 275 (July 1996): 72–78.

Fields, Scott. "Dead Again." *Earth* 4 (April 95): 16.

Kerr, Richard A. "Biggest Extinction Looks Catastrophic." *Science* 280 (May 15, 1998): 1007.

———. "Life's Winners Keep Their Poise in Tough Times." *Science* 278 (November 21, 1997): 1403.

Monastersky, Richard. "Death Swept Earth at End of Permian." *Science News* 153 (May 16, 1998): 308.

Murphy, J. Brendan, and R. Damian Nance. "Mountain Belts and the Supercontinent Cycle." *Scientific American* 266 (April 1992): 84–91.

Pendick, Daniel. "The Greatest Catastrophe." *Earth* 6 (February 1997): 34–35.

Raup, David M. "Biological Extinction in Earth History." *Science* 231 (March 28, 1986): 1528–1533.

Shell, Ellen Ruppel. "Waves of Creation." *Discover* 14 (May 1993): 54–61.

Waters, Tom. "Death by Seltzer." *Discover* 18 (January 1997): 54–55.

TRIASSIC DINOSAURS

Benton, Michael J. "Late Triassic Extinctions and the Origin of the Dinosaurs." *Science* 260 (May 7, 1993): 769–770.

Buffetaut, Eric, and Rucha Ingavat. "The Mesozoic Vertebrates of Thailand." *Scientific American* 253 (August 1985): 80–87.

Burgin, Toni, et al. "The Fossils of Monte San Giorgio." *Scientific American* 260 (June 1989): 74–81.

Fischman, Joshua. "Paleontologists Examine Old Bones and New Interpretations." *Science* 262 (November 5, 1993): 845.

Monastersky, Richard. "Eruptions Cleared Path for Dinosaurs." *Science News* 146 (July 16, 1994): 38.

Morell, Virginia. "Announcing the Birth of a Heresy." *Discover* 8 (March 1987): 26–50.

Newton, Cathryn R. "Significance of 'Tethyan' Fossils in the American Cordillera." *Scientific American* 242 (October 21, 1988): 385–390.

Padian, Kevin. "Triassic-Jurassic Extinctions." *Science* 241 (September 9, 1988): 1358–1360.

Stokstad, Erik. "Tooth Theory Revises History of Mammals." *Science* 291 (January 5, 2001): 26.

Svitil, Kathy A. "The Crystal-Bearing River." *Discover* 18 (January 1997): 26.

Wright, Karen. "What the Dinosaurs Left Us." *Discover* 17 (June 1996): 59–65.

JURASSIC BIRDS

Anderson, Alun. "Early Bird Threatens *Archaeopteryx*'s Perch." *Science* 253 (July 5, 1991): 35.

Bonatti, Enrico. "The Rifting of Continents." *Scientific American* 256 (March 1987): 97–103.

DiChristina, Mariette. "The Dinosaur Hunter." *Popular Science* (September 1996): 41–46.

Fischman, Joshua. "Were Dinos Cold-Blooded After All? The Nose Knows." *Science* 270 (November 3, 1995): 735–736.

Gurnis, Michael. "Sculpting the Earth from the Inside Out." *Scientific American* 264 (March 2001): 40–47.

Hecht, Jeff. "Contenders for the Crown." *Earth* 7 (February 1998): 16–17.

Miller, Mary K. "Tracking Pterosaurs." *Earth* 6 (October 1997): 20–21.

Morell, Virginia. "Warm-Blooded Dino Debate Blows Hot and Cold." *Science* 265 (July 8, 1994): 188.

Motani, Ryosuke. "Rulers of the Jurassic Seas." *Scientific American* 283 (December 2000): 52–58.

Padian, Kevin, and Luis M. Chiappe. "The Origin of Birds and Their Flight." *Scientific American* 278 (February 1998): 38–47.

Robbins, Jim. "The Real Jurassic Park." *Discover* 12 (March 1991): 52–59.

Thomas, David. "Tracking a Dinosaur Attack." *Scientific American* 277 (December 1997): 74–79.

Wellnhofer, Peter. "*Archaeopteryx*." *Scientific American* 262 (May 1990): 70–77.

CRETACEOUS CORALS

Alvarez, Walter, and Frank Asaro. "An Extraterrestrial Impact." *Scientific American* 263 (October 1990): 78–84.

Bird, Peter. "Formation of the Rocky Mountains, Western United States: A Continuum Computer Model." *Science* 239 (March 25, 1988): 1501–1507.

Courtillot, Vincent E. "A Volcanic Eruption." *Scientific American* 263 (October 1990): 85–92.

Gould, Stephen J. "An Asteroid to Die for." *Discover* 10 (October 1989): 60–65.

Gurnis, Michael. "Sculpting the Earth from Inside Out." *Scientific American* 284 (March 2001): 40–47.

Hallam, Anthony. "End-Cretaceous Mass Extinction Event: Argument for Terrestrial Causation." *Science* 238 (November 27, 1987): 1237–1241.

Hildebrand, Alan R., and William V. Boynton. "Cretaceous Ground Zero." *Natural History* (June 1991): 47–52.

Labandeira, Conrad C. "How Old Is the Flower and the Fly?" *Science* 280 (April 3, 1998): 57–59.

Larson, Roger L. "The Mid-Cretaceous Superplume Episode." *Scientific American* 272 (February 1995): 82–86.

Monastersky, Richard. "Closing in on the Killer." *Science News* 141 (January 25, 1992): 56–58.

Morell, Virginia. "How Lethal Was the K-T Impact?" *Science* 261 (September 17, 1993): 1518–1519.

Novacek, Michael J., et al. "Fossils of the Flaming Cliffs." *Scientific American* 271 (December 1994): 60–69.

Vickers-Rich, Patricia, and Thomas Hewitt Rich. "Polar Dinosaurs of Australia." *Scientific American* 269 (July 1993): 49–55.

TERTIARY MAMMALS

Culotta, Elizabeth. "Ninety Ways to Be a Mammal." *Science* 266 (November 18, 1994): 1161.

Monastersky, Richard. "The Whale's Tale." *Science News* 156 (November 6, 1999): 296–298.

Pendick, Daniel. "The Mammal Mother Lode." *Earth* 4 (April 1995): 20–23.

Perkins, Sid. "The Making of a Grand Canyon." *Science News* 158 (September 30, 2000): 218–220.

Pinter, Nicholas, and Mark T. Brandon. "How Erosion Builds Mountains." *Scientific American* 276 (April 1997): 74–79.

Ruddiman, William F., and John E. Kutzbach. "Plateau Uplift and Climate Change." *Scientific American* 264 (March 1991): 66–74.

Schmidt, Karen. "Rise of the Mammals." *Earth* 5 (October 1996): 20–21 and 68–69.

Schueller, Gretel. "Mammal 'Missing Link' Found." *Earth* 7 (April 1998): 9.

Simpson, Sarah, "Wrong Place, Wrong Time, Right Mammal." *Earth* 7 (April 1998): 22.

Stokstad, Erik. "Exquisite Chinese Fossils Add New Pages to Book of Life." *Science* 291 (January 12, 2001): 232–236.

Storch, Gerhard. "The Mammals of Island Europe." *Scientific American* 266 (February 1992): 64–69.

Sullivant, Rosemary. "Flapping Through the Bottleneck." *Earth* 6 (December 1997): 22–23.

Watson, Andrew. "Will Fossils From Down Under Upend Mammal Evolution?" *Science* 278 (November 21, 1997): 1401.

Zimmer, Carl. "Fossil Offers a Glimpse into Mammal's Past." *Science* 283 (March 26, 1999): 1989–1990.

QUATERNARY GLACIATION

Allen, Joseph Baneth, and Tom Waters. "The Great Northern Ice Sheet." *Earth* 4 (February 1995): 12–13.

Broecker, Wallace S., and George H. Denton. "What Drives Glacial Cycles." *Scientific American* 262 (January 1990): 49–56.

Flannery, Timothy F. "Debating Extinction." *Science* 283 (January 8, 1999): 182–183.

———. "The Case of the Missing Meat Eaters." *Natural History* 102 (June 1993): 41–44.

Kerr, Richard A. "Slide Into Ice Ages Not Carbon Dioxide's Fault." *Science* 284 (June 11, 1999): 1743–1746.

Kimber, Robert. "A Glacier's Gift." *Audubon* 95 (May–June 1993): 52–53.

Leakey, Meave, and Alan Walker. "Early Hominid Fossils from Africa." *Scientific American* 276 (June 1997): 74–79.

Matthews, Samuel W. "Ice on the World." *National Geographic* 171 (January 1987): 84–103.

Monastersky, Richard. "Stones Crush Standard Ice History." *Science News* 145 (January 1, 1994): 4.

Rodbell, Donald T. "The Younger Dryas: Cold, Cold Everywhere?" *Science* 290 (October 13, 2000): 285–286.

Tattersall, Ian. "Once We Were Not Alone." *Scientific American* 282 (January 2000): 56–62.

Waters, Tom. "A Glacier Was Here." *Earth* 4 (February 1995): 58–60.

INDEX

Boldface page numbers indicate extensive treatment of a topic. *Italic* page numbers indicate illustrations or captions. Page numbers followed by *m* indicate maps; *t* indicate tables; *g* indicate glossary.